CW00727519

MODERN ELECTRONIC TEST E

Modern Electronic Test Equipment

Second Edition

Keith Brindley

Heinemann Newnes

To Thomas Alexander, the second!

Heinemann Newnes
An imprint of Heinemann Professional Publishing Ltd
Halley Court, Jordan Hill, Oxford OX2 8EJ

OXFORD LONDON MELBOURNE AUCKLAND
SINGAPORE IBADAN NAIROBI GABORONE
KINGSTON

First published 1986
Second edition 1990

© Keith Brindley 1986, 1990

British Library Cataloguing in Publication Data
Brindley, Keith
Modern electronic test equipment. 2nd ed.
1. Electronic testing equipment
I. Title
621.381548

ISBN 0 434 90063 X

Typeset by Hope Services (Abingdon) Ltd
Printed by Redwood Press, Melksham, Wiltshire

Contents

Preface

Remarkably, it has been four years since the first edition of *Modern Electronic Test Equipment*. A great deal has happened in the world of test equipment during those years, particularly in the areas associated with automatic and optical fibre test equipment. This new edition represents a substantial revision, incorporating these and many other changes. Indeed, the book has almost doubled in content.

Main aims of the first edition have not been compromised. My idea now, as then, brings together the varieties of instruments to discuss them in a down-to-earth and generalised, but nevertheless, in-depth fashion. With this in mind and without being concerned with precise circuit details of individual instruments and techniques, Part One of this new edition discusses typical design principles commonly used in modern electronic test equipment. New or substantially revised chapters are included on newly developed aspects and equipment; time domain reflectometers, automatic test equipment, instrument buses (particularly VXIbus – a method of interfacing modular test instruments which, quite literally, is set to revolutionise modern electronic test equipment).

Part Two also covers new ground, discussing techniques and methods of measurement, again without being concerned with individual instruments. This is essentially an alphabetic list of measurands, with suggested methods of measurement. It is designed for simplicity and ease of use – other reference sources show ways instruments are used to measure measurands so, if a reader wants to find out different methods of measuring a particular measurand, each instrument must be studied first to ensure all methods have been found. An alphabetic list of measurands, on the other hand, as found in Part Two of this book makes it far easier to locate an acceptable method.

Who will benefit from the book? In short, anyone who comes into contact with electronic test equipment. This is a practical book; not in the way that it shows you how to *use* test equipment (it doesn't) but in that it tells how the equipment works and what it is capable of. Knowing this, any engineer, technician, scientist, student, teacher, graduate, manager and so on will be more able to understand the equipment at hand. Moreover, anyone in a purchasing position needs to know whether the test equipment recommended will be able to do the job it's required to do. Knowledge gained here will make the decision easier.

Keith Brindley

Part One Equipment and Principles

1 Introduction

Test equipment is used in all technical areas. Whether in the manufacture of electronic appliances (and all the design, development, production, test, QA, etc., which that implies), or in the servicing of appliances, or in education, or in the sciences, test equipment is required simply to observe how the appliance operates.

The past

In the 'good old days' the engineer's complete range of test equipment instruments usually comprised an Avometer, an oscilloscope, and perhaps a signal generator of some description. With these, most appliances of the day could be manufactured, serviced, or studied quite adequately. The electronics world was virtually all analog and the analog test instruments available did their jobs just fine.

The present

Nowadays, things have changed. Appliances are often digital in nature, microprocessor-based in technique, and far more complex in operation. They generally perform better, however, and – with exceptions – are more reliable.

But more complex appliances require correspondingly more complex test equipment. And not only is the test equipment more complex, but new *types* of test instrument are generated too. The most obvious example of a new generation of test instruments in recent years is the logic analyser, which was developed in response to the basic need to observe internal operations of micro processor-based systems. Prior to the microprocessor, testing an appliance was often as simple as applying a suitable signal at the appliance's input and observing signals obtained at each stage throughout the appliance – one at a time. When the microprocessor bus came along, it brought with it the requirement to observe a great deal more than one signal – and all of these, simultaneously. And so the logic analyser was born. Doubtless, further generations of test instruments will be spawned by future appliance needs.

The future

As test equipment becomes more complex to cope with the increased demands of more complex appliances – especially those of a microprocessor-based nature – it is correspondingly natural that the test equipment itself becomes microprocessor-based. A large number of microprocessor-controlled meters, oscilloscopes, counters, signal generators, etc. have already been manufactured and almost every new test instrument coming onto the market has that distinction. Microprocessor control allows instruments to have features which were not previously possible. Typically, a microprocessor-controlled test instrument has a 'soft panel', that is, a keypad on which to enter the data required to configure the instrument for the measurement task; previously this would be undertaken with hardwired mechanical switches. Microprocessor-controlled test instruments with cathode ray tube displays, apart from the pure analog information which older instruments are only capable of, often have alphanumeric data presented on-screen, too.

But features like these, and the present state-of-the-art in microprocessor-controlled test equipment, are still nothing to what is around the corner. Looking again at the logic analyser, a new type of test instrument, we can see that it is, to all intents and purposes, just a microprocessor-based computer. It has, however, input and output functions which correspond directly to its specific task as a logic analyser, and it is internally programmed to perform this task. So, although it is a computer system, it can be used for no other purpose than logic analysis.

This begs a question. Why can't a logic analyser be reprogrammed for other tasks, say spectrum analysis, or voltage/current/resistance measurements, or counting, or even as a signal generator? A computer, after all, is a prime example of deferred design: it is manufactured as a multifunctional tool, and the decision about its eventual operational task is deferred until a program is executed. A computer is the ultimate piece of designer jewellery – you just can't wear it round your neck, that's all!

In this respect, future generations of test equipment may be universal: capable of all forms of test and measurement, at the push of a soft-panel button. We're already part way there. Presently available automatic test equipment is capable of this in a limited way. The user can program the automatic test equipment system to perform all kinds of tests and measurements, provided the corresponding system parts are there. But this is not *universal test equipment* (my term) in the sense given above. It is not one computer performing all tasks merely by a change in the controlling program, for an automatic test equipment system is more a collection of computers (one for each task) controlled in the way the user requires.

It may be that automatic test equipment systems are the closest we will get to universal test equipment of my description. I am not a gambler. However, I am prepared to stake a bottle of good wine in favour of universal test equipment. I am sure I have backed a winner. Cheers!

Explanation of terms

It is useful to have a look at some of the basic and important technical terms used throughout the book. These terms are all related and require rigid definitions before they are used, in order that they are not *abused*. They are by no means the only important terms with respect to test equipment, but are the main ones.

Often, writers (particularly our North American cousins) use terms such as *system under test* (SUT) or *unit under test* (UUT) to describe both the appliance being tested and the measurements taken on the appliance. I can't say that I like these terms and I'm certainly not going to use them in such an indiscriminate way. As it is a specific function of an appliance which is always measured and displayed by test equipment and not the appliance itself (the measurements – height, width, depth etc. – are usually given in the appliance specification and don't change, so they don't often need to be re-measured), I shall refer to the quantity being measured as the **measurand:** a term regularly used in the field of study of electronic instrumentation. With respect to the test equipment discussed in this book, it's probably true to say that in most applications the measurand being measured and displayed is an electrical quantity: current, voltage, or frequency. However, certain applications may arise in which the measurand is a *physical* quantity such as strain, force, displacement, velocity etc. In such cases, transducers are generally used to convert the physical measurand into an electrical one.

The **accuracy** of a measurement, that is, the closeness of the measured value of a measurand to its actual value, is generally specified in the terms of **error:** the maximum possible difference between measured and actual values. For example, a 300 mm rule may have an error of, say, ± 1 mm. This means that the rule itself may have an actual length of somewhere between 299 mm and 301 mm – it *may* be exactly 300 mm, on the other hand it might not! Any measurement taken with the rule therefore has a maximum possible error of 1 mm, high or low. Sometimes, error is specified as a percentage. In the case of the rule, error could be specified as ± 0.0033 per cent. Sometimes, in special cases, the error is specified as a percentage of **full scale deflection,** that is, as a percentage of the maximum reading. Errors can be a factor of the test instrument used, or can be user-generated.

The fineness with which a measurement can be taken is known as the **resolution.** If the rule is graduated in millimetres, then it should be possible to interpolate between two millimetre markings when measuring, to give a resolution of 0.5 mm. However, the fact that the resolution may be lower than the specified error does not mean that the reading has a lower error; an important point. The overall error is, in fact, greater.

Book layout and description

There is a great diversity of instruments classed as test equipment. Some of these instruments are extremely specialised, and suit only one or two applications. Their inclusion in a book like this would be pointless: it is not my intention to attempt to cover all types of test equipment – that is impossible. Fortunately, broad categories of test equipment occur quite naturally. Correspondingly, the book covers the main categories, chapter by chapter.

Categories discussed are those of *modern* test equipment; or, more specifically, test equipment which is most commonly used in modern electronics labs. For this reason, and at risk of annoying some readers, I have not detailed to any depth test equipment, e.g. bridge instruments, which I do not consider to be modern. However, in a slight contradiction, I *have* included the category of analog meters, arguably a dying breed of test equipment instruments. This is because most, if not all, basic principles of tests and measurements can be usefully studied by such a discussion. I don't think any other category of test instruments allows this to the same extent.

The advent of microprocessor-based test equipment presented me with an unusual predicament when trying to describe individual instruments. I have used block diagrams wherever possible to illustrate operational explanations, and in the case of the common and, dare I say it, old-fashioned instruments, say, oscilloscopes or signal generators, there is no problem. But where these instruments now have a microprocessor-based counterpart the block diagrams cannot *strictly* correspond – the microprocessor-based counterparts are, after all, computer systems, and don't necessarily operate in the same ways as before. Nevertheless, the block diagram illustrations given throughout the book stand: albeit in a more esoteric way. Block diagrams of this sort cannot be readily given of the new types of instrument, however, which could not be made in a non-microprocessor-based way. So simple diagrams of a microprocessor-based computer architecture have to suffice.

Finally, the book is not a guide on which test equipment instruments to buy. It is a reference book of how the main categories of test equipment work, allowing the reader: (1) to compare available instruments and make an informed choice; and (2) to use the equipment to the best advantage.

2
Analog meters

The most common test instrument encountered in the laboratory is the analog meter, using the moving-coil movement as its display. This is a seemingly odd state of affairs as the analog meter is not, as we shall see, particularly accurate. However, it is a versatile and general-purpose tool which is used in many applications to give reasonable results. We use it here as a vehicle which allows discussion of some major points concerning test equipment.

Moving-coil movements of today (discussed in Appendix 1) represent the culmination of developments, stretching back over 150 years, since the first discovery that a current passing through a wire suspended in a magnetic field created a force which tended to move the wire. In the moving-coil movement, the wire is coiled and rotates in proportion to the current, turning, and taking with it a pointer which points to a graduated scale, indicating the value of the current. Its extreme simplicity, yet comparative accuracy, means that the moving-coil movement forms the basis of many items of test equipment. Indeed, until just a few years ago, most measurements taken in electrical and electronic circuits were effected using the moving-coil analog display. Today, things are beginning to change, and although still a very popular instrument, the analog meter is slowly being overtaken by its digital counterparts. It will be some years yet, however, before the moving-coil movement itself is made obsolete, if ever; because in some measurements the movement gives a better display than is possible in available digital devices.

General-purpose meters

Moving-coil movements are generally manufactured with **sensitivities** ranging between about 10 μA and 1 mA at **full scale deflection** (FSD); i.e. when the pointer is deflected to the furthermost point by the current. The coil within the movement has resistance, of course, ranging in value between about 5 and 5000 Ω, which from Ohm's law means that voltage values of between about 50 μV and 5 V will deflect the pointer to full scale deflection. Any one coil will only be able to measure one particular range of values of current or voltage, of course, so if the movement is to be used to measure a number of ranges of values, some way must be incorporated of converting the measured value to within the movement's own range. In the simplest form, the measured converter is a single resistor connected either in series or in parallel with the movement. When in series, the converter (known as a **multiplier**) effectively raises the total resistance, allowing the combination of movement and multiplier to measure a higher voltage. When in parallel, the converter (known now as a **shunt**) effectively lowers the total resistance, allowing the combination to measure higher currents. Note that in both cases no more current passes through the movement's coil than before: only sufficient current to produce full scale deflection of the pointer.

With an external power source such as a cell, the movement can also be used to indicate values of resistance. By connecting the movement in series with a cell and the resistance to be measured the pointer will indicate the current which passes through the resistor. As this current, from Ohm's law, is *inversely* proportional to the resistance, the indication on the scale of the movement will *also* be inversely proportional to the resistance. In effect, the scale graduated for resistance must be marked in the

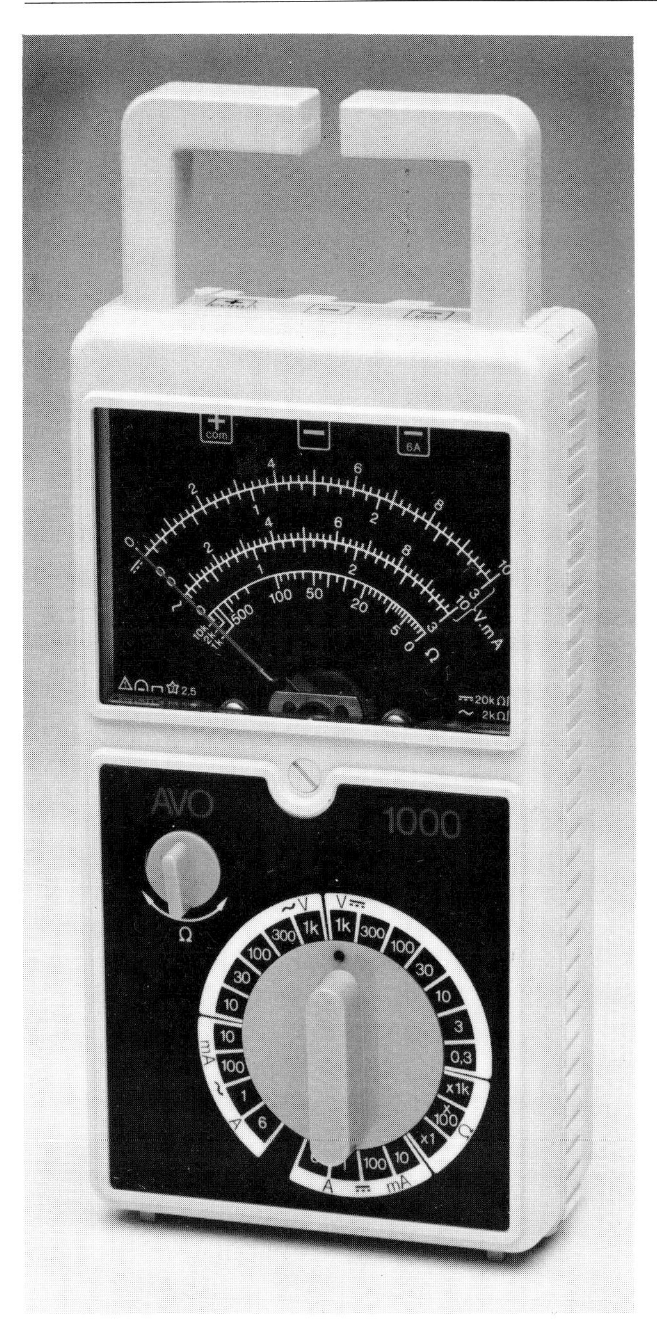

Photo 2.1
Avometer 1000 analog multimeter (Thorn EMI Instruments)

opposite direction to current and voltage scales, that is, the further up the scale the pointer is deflected, the lower the resistance under measurement.

Using mechanical switches to switch multipliers, shunts and cells in and out of circuit with a moving-coil movement, an analog meter capable of measurement of many ranges of voltage, current and resistance can be constructed. Such an instrument earns the name **volt-ohm-milliammeter** (VOM), although more commonly it is called a **multimeter.** The basic multimeter is shown in block diagram form in Figure 2. 1.

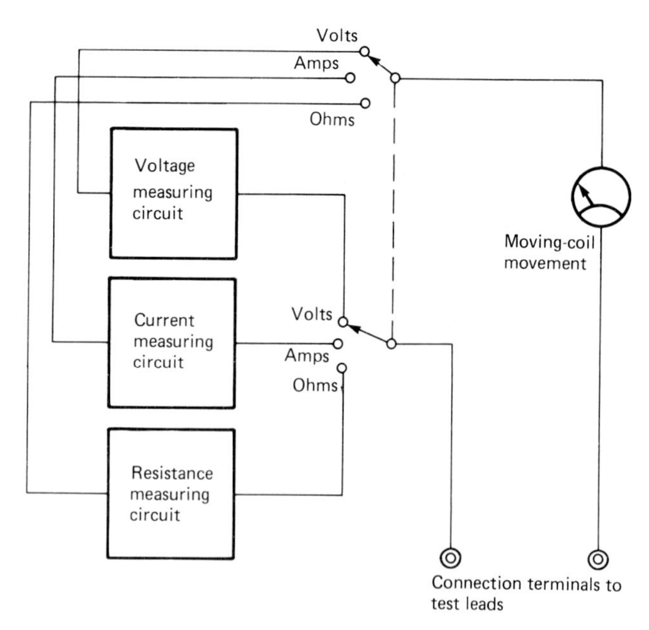

Figure 2.1 Block diagram of a basic analog multimeter

DC voltage measurements

A typical configuration of multipliers and switch is shown in Figure 2.2 for a 50 μA meter with a coil resistance of 2000 Ω. Six ranges are available, five of which, 2.5 V, 10 V, 50 V, 250 V, 1000 V, are switched

Photo 2.2 Avometer Model 8 mark 6 analog multimeter (Thorn EMI Instruments)

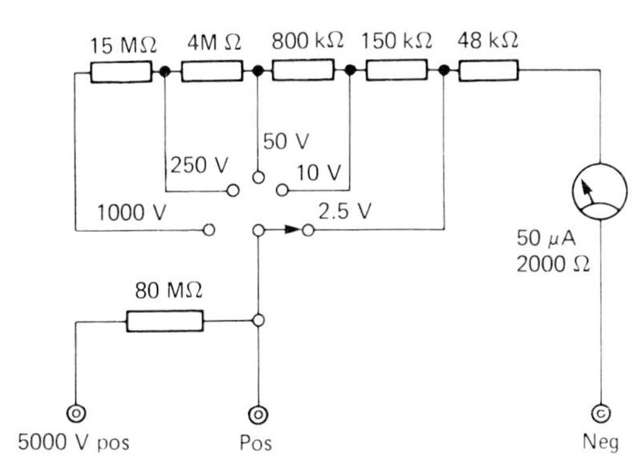

Figure 2.2 Typical configuration of multipliers and switch in an analog multimeter, for DC voltage measurement

into circuit via the mechanical rotary switch, while the sixth range, 5000 V, is selected by the user by connecting the positive probe into a separate connection socket which has an 80 MΩ resistance in series. A separate socket is needed because typical rotary switches cannot withstand voltages over 1000 V. With the switch in the 2.5 V range position it's easy to calculate that the total series resistance is 50 kΩ (that is, 48 kΩ + 2 kΩ), so the input resistance (see later) of the meter is 50 000/2.5, or 20 000 Ω/V. Similar calculations show this to be so on each range.

DC current measurement

Figure 2.3 shows a moving-coil movement, shunts and switch connected into a DC current measuring circuit. This particular meter circuit is known as a **ring-shunt** configuration. The ring-shunt's main advantage is that the movement is always shunted, even when the range switch is turned from one position to another, thus protecting it from accidental burnout, even if the test leads are connected to a 'live' circuit.

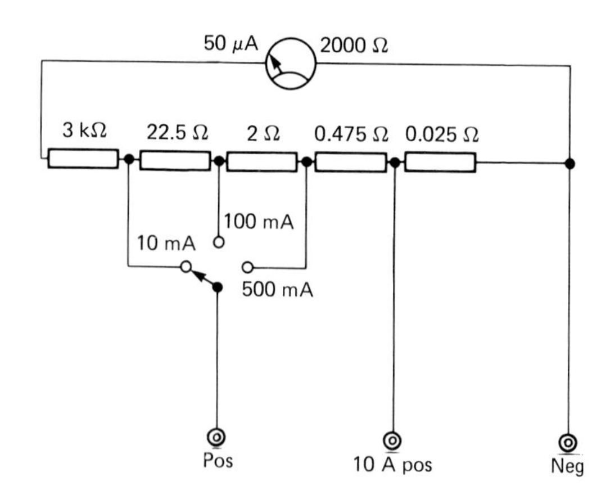

Figure 2.3 Typical configuration of shunts and switch in an analog multimeter, for DC current measurement

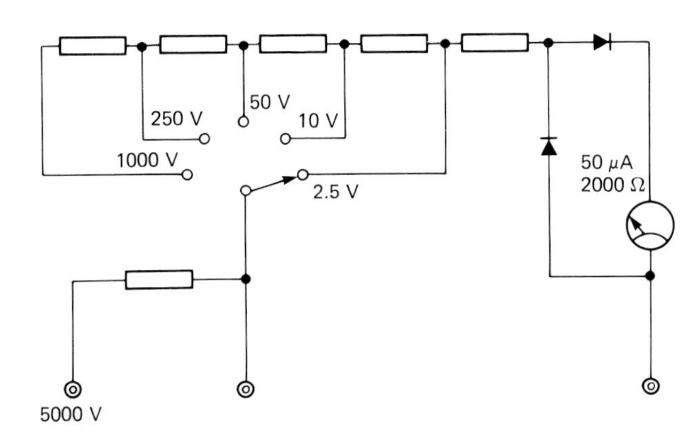

Figure 2.4 Typical configuration of multipliers, switch and diodes in an analog multimeter, for AC voltage measurement

AC voltage measurements

A circuit for a typical AC voltage measurement is shown in Figure 2.4. It is similar to the DC voltage measurement circuit of Figure 2.2, but the input resistance is lower, say one-third the DC circuit's. This is mainly because the applied AC voltage must be rectified by a diode rectifier before application to the meter. The meter movement responds to the average value of the rectified AC voltage, that is, 0.318 of the peak value, for a half-wave rectifier as in this example. In turn, therefore, the input resistance of the AC voltmeter circuit is reduced to 31.8 per cent of its DC value, or 6360Ω/V. Interestingly, if a multimeter using this basic rectification technique is used to measure a DC voltage while set to an AC range, a reading much higher than expected would be displayed. This is because the voltage scales are calibrated to indicate RMS values of sine waves, that is 0.707 of peak value. So the indicated voltage, E_{ind}, is:

$$\frac{\text{RMS voltage}}{\text{Average voltage}} = \frac{0.707}{0.318} = 2.22 \text{ times the DC voltage.}$$

Similarly, as the average value of a full-wave rectified voltage is 0.636 of the peak value, a meter using a full-wave bridge rectifier will indicate 1.11 times the DC voltage.

Circuits like this aren't particularly accurate because they rely on the AC measurand being of a pure sinusoidal form. Any small distortion in the measurand causes a corresponding inaccuracy. Also, circuit inductances and capacitances limit the upper frequency response of basic AC meter circuits to around 10 kHz. Low frequency response goes down to 0 Hz, but a practical limit is about 10 Hz, as pointer vibration becomes objectionable at lower frequencies of measurand. They are, nevertheless, used in general-purpose analog multimeters for their cheapness, sensitivity, and fast measurement speed.

Supplementary range functions for specialised applications are often provided in general-purpose multimeters of this type by external or internal multipliers, shunts, probe arrangements, transformers, transducers and other devices, but the basic analog meter arrangement is no more than shown and described here.

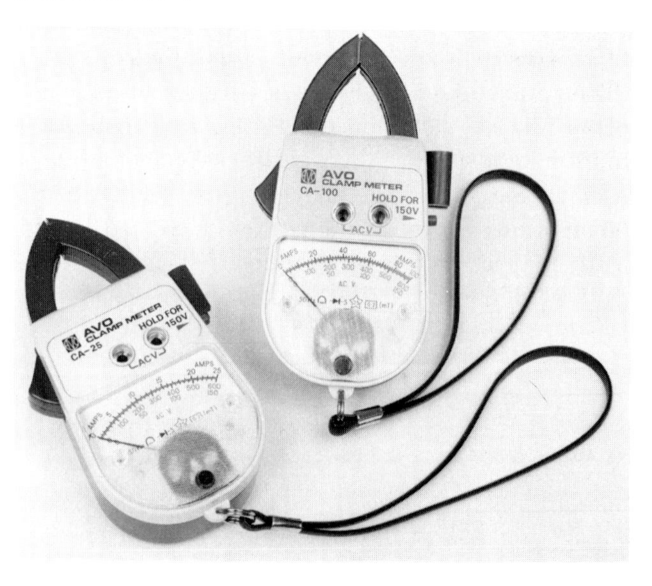

Photo 2.3 Avo analog clampmeters, models CA-25 and CA-100 used for measuring alternating current without the need to break the circuit under test (Thorn EMI Instruments)

Sources of error

Multimeters are designed as utility instruments and, as such, have a moderate accuracy of around ± 2 to ± 5 per cent. Higher quality analog meters have correspondingly better accuracies. Provided that allowance is made for the loading effects which the instrument may have on the measurand circuit, this is more than adequate for most measuring needs.

The accuracy quoted for a multimeter is specified as a percentage of full scale deflection, which means that the accuracy is dependent on the position of the pointer on the scale for any particular measurement. For example, a meter specified as having an accuracy of ± 4 per cent only, has an accuracy of ± 40 per cent if the reading is taken with the pointer at one-tenth of full scale deflection. To avoid high inaccuracies a multimeter would therefore normally be used to take measurements where the readings lie in the upper third of the scale, by having scale ranges which overlap, in 1–2–5 or 1–3–10 ratios. A possible multimeter scale combination is shown in Figure 2.5. This combination of careful use and good design helps to restrict reading errors to a respectable level.

The input resistance of a multimeter on its DC voltage measurement function ranges is typically around 20 000 Ω/V. This means that the meter on, say, its 10 V range, has an input resistance of 200 000 Ω and on its 0.1 V range has an input

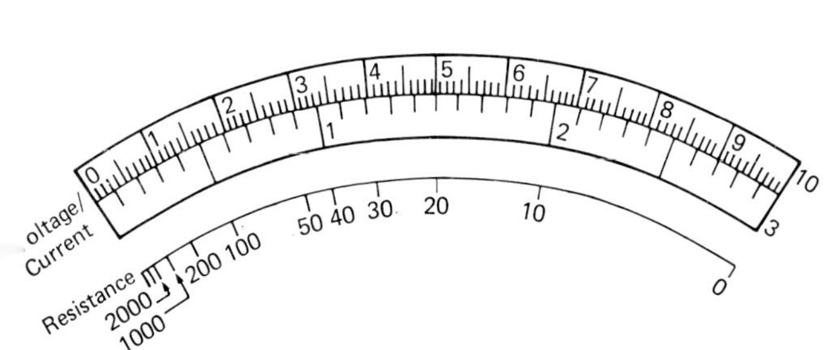

Figure 2.5
Typical analog multimeter scale configuration

resistance of 2000 Ω. Because the effects of loading on the measurand circuit increase as the input resistance of the meter decreases, the user must bear in mind that the accuracy of the measured reading will depend on the meter's resistance. As a rule of thumb, the meter's input resistance should be at least (and preferably more than) ten times the resistance of the measured circuit at the measured points. Less than this and the meter will load the circuit, causing inaccurate readings (which may sometimes be compensated for, as shown below). If the measurand circuit's resistance is unknown, a test can be carried out to find if the circuit is loaded, and if so a correction can then be made to the readings to calculate the actual voltage. The test is straightforward: take meter readings of the measured voltage on two of the meter's voltage ranges. If the measurand circuit has a resistance which is high enough to cause meter loading, the two readings will differ substantially. The actual voltage can now be calculated from the formula:

$$V = \frac{E_1 E_2 (R_2 - R_1)}{E_1 R_2 - E_2 R_1}$$

where: V = true voltage
E_1 = meter reading on first range setting
E_2 = meter reading on second range setting
R_1 = input resistance of meter on first range setting
R_2 = input resistance of meter on second range setting.

Strictly, this correction method is usable only on linear systems. However, it can be used in most instances, where a meter of high enough input resistance is not available, to give, at least, a more accurate measurement than would otherwise be possible.

Another factor which may affect the accuracy of a measurement taken with an analog meter is the resolution with which the measurement is perceived by the user. Strictly speaking, of course, this is a user-error and is not the fault of the instrument. There are two main errors possible here: first, that of parallax; and second, misjudgement as to the exact position of the pointer.

Specialised analog meters

Where basic passive moving-coil movements do not give a high enough meter input resistance to give negligible loading effects, or where the correction method above can't be used, it is possible to use active circuits in combination with the movement to increase resistance. Current amplifiers are used in such multimeters, based around an op-amp or field effect transistor circuit, having input resistances in the order of megohms. These types of multimeters also allow more accurate measurement of AC measurands, because circuits can be included which allow true RMS readings, even if the measurand is not a sinewave, up to frequencies of about 100 MHz. Multimeters with these **true RMS converter** circuits are given a **crest factor** specification, which defines the types of waveforms which can be measured accurately – see Chapter 3 for details.

The inclusion of active circuits along with the meter movement need not stop with the result of just a multimeter. Other types of analog meter may be constructed, using the standard moving-coil movement as the display device, but with the extra active circuits which define the meter's particular function.

Photo 2.4 Racal-Dana 9103 absorption RF wattmeter (Racal Group Services)

Analog power measurements

A good example of how this is so can be seen in the case of a **power meter** which uses thermal measuring techniques. In such a power meter the measurand whose power is to be measured is applied on an internal resistive load. A thermocouple, or similar device, measures the temperature of the load relative to ambient temperature. The output voltage of the thermocouple is amplified and drives a meter movement which is calibrated to read power. Figure 2.6 shows the principle as a block diagram.

Figure 2.6 Block diagram of a thermal measurement power meter

Most power meters are marked not only in milliwatts and watts, but in decibels also, and, generally, the 0 dB mark will coincide with the 1 mW mark on the meter's scale. A dB scale means that the power meter can now be used to measure frequency response and signal-to-noise ratios. Frequency response of a measurand circuit is most easily measured by applying an input signal to the circuit in frequency steps and measuring the output signal at those steps. The frequency response is defined as the band of frequencies within the two corner frequencies, where the corner frequencies are marked by the frequencies at which the circuit's output power falls by 3 dB below the mid-band output power. Signal-to-noise ratios, on the other hand, are easily measured by first measuring the output power of the measurand circuit with an applied signal, then measuring again with no signal. The change in power levels is the signal-to-noise ratio.

Distortion measurements

Analog meters are often used to measure distortion. Generally, the instruments used to do this are known as **distortion analysers**; they are, however, incorrectly named, because they do not *analyse* the distortion in the true sense of the word, they merely indicate the level of distortion present.

Distortion is a general term referring to the changes impressed on a signal as it passes through a system. Thus, if a system's output signal is different from the input signal in any way, it is distorted. In this sense any difference, such as frequency response changes, phase shifts, or even noise, is a distortion. More specifically, on the other hand, the types of distortion we are going to consider here usually refer to the addition of extra frequency components, which are related to the input signal frequency in some way. The most common types of distortion which would need to be measured are **harmonic distortion** and a number of varieties of **intermodulation distortion.**

Harmonic distortion occurs where the extra frequency components added to the signal are harmonics of the signal. So, if an input signal of frequency f is applied to a system, the harmonically distorted output signal might contain frequency

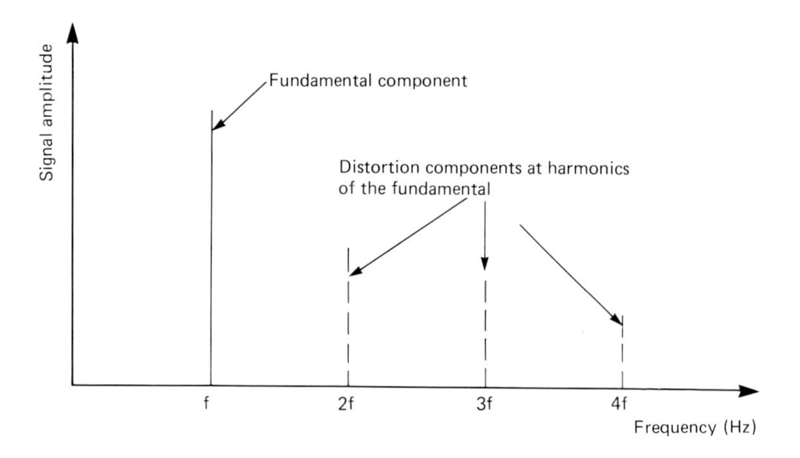

Figure 2.7 Frequency components of a harmonically distorted signal

components of the frequencies f, $2f$, $3f$, $4f$ etc. Shown as a graph of signal amplitude against frequency, as in Figure 2.7, the additional frequency components of the harmonically distorted signal are seen to be integral multiples of the fundamental frequency. The size of these additional frequency components, relative to the fundamental frequency component, determines the amount of harmonic distortion present. The measurement of harmonic distortion is in fact the ratio, expressed as a percentage, of the sum of all of the harmonics compared to the fundamental frequency component.

Figure 2.8 shows a block diagram of a harmonic distortion measurement system, which may be used in distortion analyser. The signal converter block converts the applied signal to a reference level, typically 1 V, so a combination of attenuating resistors and an amplifier would be used. Thus, signals larger than the reference level would be attenuated, signals smaller would be amplified. The reference level signal is then applied to a notch filter, which is accurately tuned to remove the fundamental frequency component thus leaving only the harmonics. After this the harmonic-only

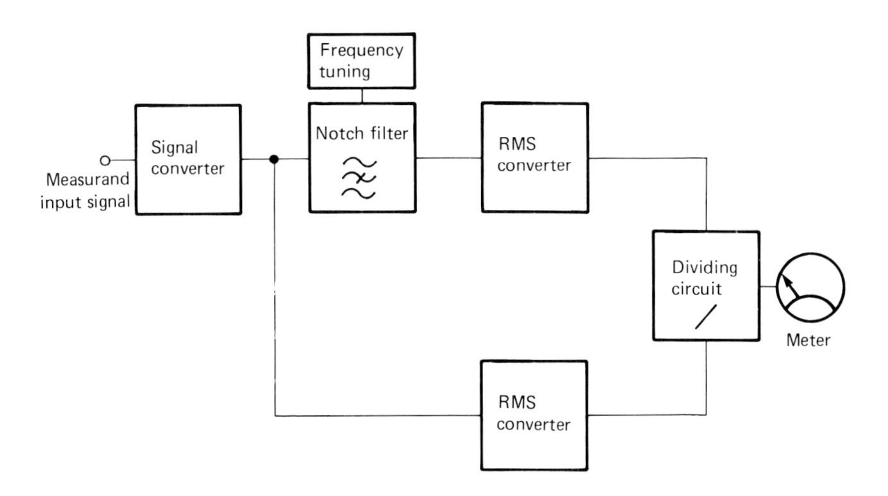

Figure 2.8 Block diagram of a harmonic distortion measurement system

signal is applied to an RMS converter. Meanwhile, the reference level signal is applied direct to another RMS converter without filtering. A dividing circuit calculates the ratio between the two RMS values and the circuit's output drives a meter, the pointer position of which indicates the level of harmonic distortion. Older methods of harmonic distortion measurement required that the input signal was first manually set to an accurately defined reference level, then the RMS level of the harmonic components was displayed and the measurement taken. This newer method renders such a two-stage measurement unnecessary.

Intermodulation distortion occurs when two or more frequency components interact in a non-linear system to produce additional components which are not harmonically related. Figure 2.9 shows a graph of amplitude against frequency for a case of intermodulation distortion where the distortion is caused by the interaction of two frequency components at frequencies of f and Xf. Distortion components may be produced at frequencies of $Xf-f$, $Xf+f$, $Xf-2f$, $Xf+2f$, etc. The signals around the frequency com-

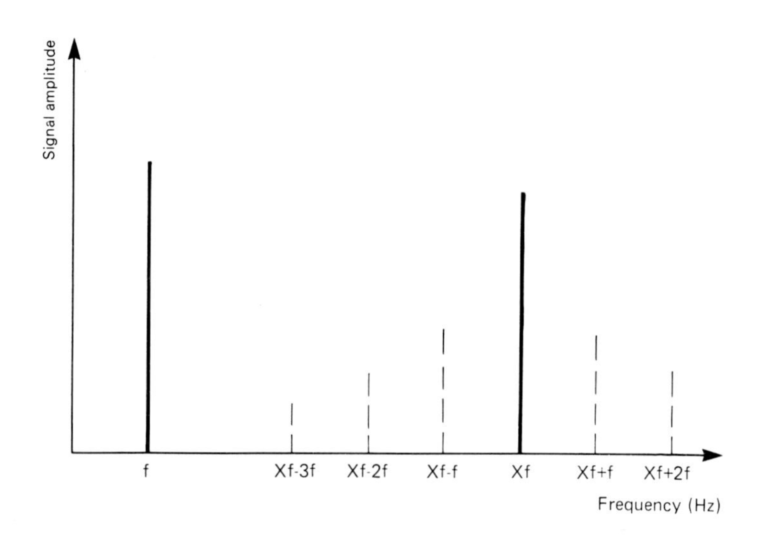

Figure 2.9 Frequency components of an intermodulation distorted signal

Figure 2.10 Block diagram of an intermodulation distortion measurement system

ponent X*f* are sidebands, identical to sidebands of a modulated radio signal.

There are other types of intermodulation distortion, each of which requires a slightly different measurement technique, but the basis of all intermodulation distortion measurements is shown, in block diagram form, in Figure 2.10. Like the harmonic distortion measurement system of Figure 2.8, the first block is a

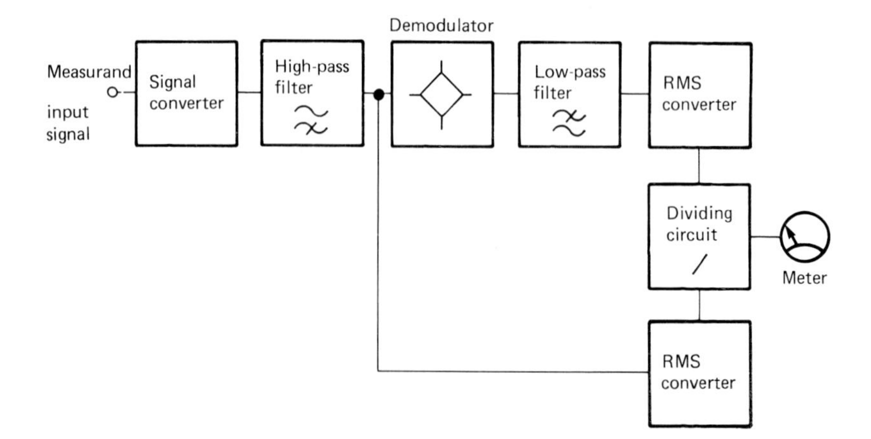

signal converter, which allows the input signal to be set to a reference level. The reference level signal is then passed through a high-pass filter which is tuned to pass all frequency component above the lower of the two interacting frequency components. A demodulator circuit separates the sidebands (at f, $2f$, etc.) from the frequency component Xf. The signal is then passed through a low-pass filter, accurately tuned to remove the Xf frequency component. The remaining distortion components are then passed through an RMS converter. Meanwhile, part of the unmodulated signal (that is, the Xf component complete with its sidebands) is passed through another RMS converter. The two signals are applied to a dividing circuit which electronically calculates the ratio between the signals, then drives the meter movement to display the result.

Modern distortion analysers are typically much more complex than these simple explanations would suggest. Top of the range systems are more-or-less fully automatic in filter tuning and level setting, and have low distortion oscillators incorporated so that the user merely has to connect the circuit to be tested to the analyser, switch on and wait for the result to be displayed.

3
Digital meters

In electronics, more and more functions which previously were regarded as exclusively analog in nature are now being accomplished using digital techniques. An obvious example is the compact disc sound recording medium, using digitally recorded discs to hold music, which on replay presents the listener with high-quality sound, free from the inherent weaknesses of the existing analog recording medium of vinyl records.

Digital techniques, although generally more complex and more expensive (initially, at least) than their analog counterparts, have a number of operational advantages: they are more accurate, more reliable and may offer increased performance. Digital meters are the epitome of this. Their internal circuits are necessarily much more complex than those of analog meters, because the nature of the measurement is such that inherently analog measurand values are converted to digital ones before display.

Accuracy of measurements taken using digital meters, on the other hand, is not limited to the accuracy of a mechanical meter movement or to the resolution with which an observer reads the display. Instead, accuracy is purely dependent on the meter's circuits, so much higher accuracy is possible.

Like analog meters, digital meters are most commonly encountered as general-purpose multimeters, although the possibilities seen in Chapter 2 of power meters, distortion meters etc., are all available using the digital meter digital multimeter as a display. The typical **digital multimeter** (DMM) performs all the usual functions of the analog meter, but generally at

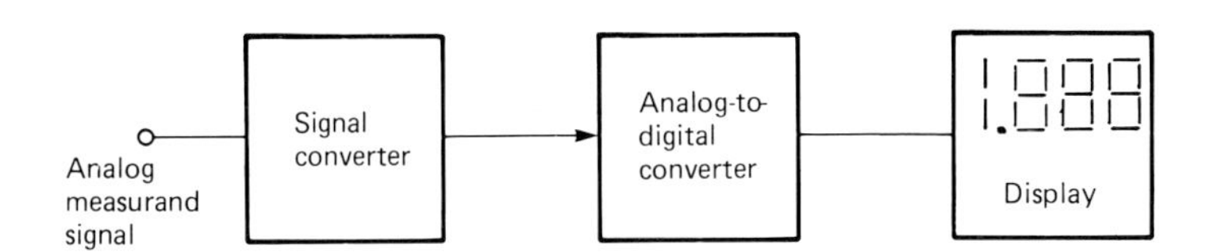

Figure 3.1 Block diagram of a basic digital multimeter

a higher performance. The block diagram of Figure 3.1 shows the action of a basic digital multimeter, albeit in a simplified form. The signal converter block converts the measurand functions of voltage, current, resistance, etc., into a form which the analog-to-digital converter block can deal with; typically 200mV DC or 2 V DC. Analog-to-digital conversion then takes place and the result is displayed digitally.

Signals and functions

Signal conversion is essentially similar to the analog multimeter's. In most digital multimeters a resistive divider chain attenuates the measurand voltage value in switched steps, so that the output voltage is within the input range of the analog-to-digital converter. AC voltage measurement is usually done with the use of a circuit called a **true RMS converter**, which measures the RMS value of applied measurand AC voltage, whether they be sinewaves or otherwise, and presents this as a DC voltage to the analog-to-digital converter. Multimeters (digital or analog) which use a true RMS converter are given a **crest factor** specification, which indicates the ability of the converter to correctly measure and display different AC waveforms. The crest factor of an AC waveform is the ratio of peak voltage to RMS voltage, so if a waveform's crest factor lies within the crest factor specification of a multimeter it will be accurately measured. Examples of waveform crest factors are: square wave, 1; sinewave, 1.414; triangular

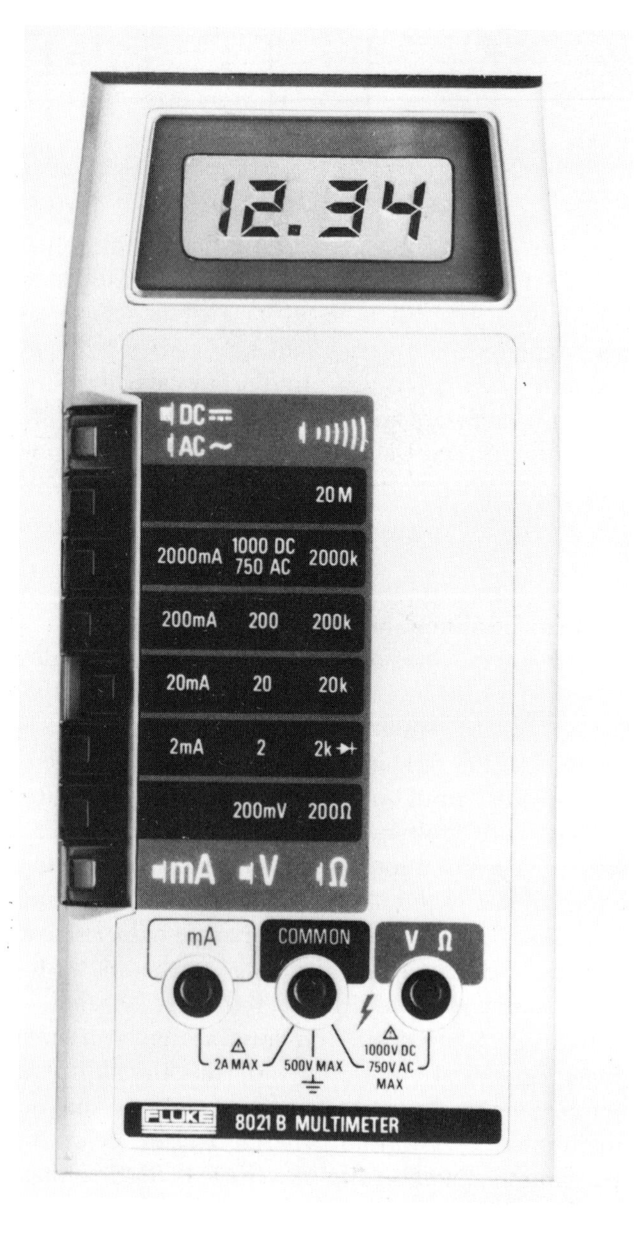

Photo 3.1 Fluke 8021B hand-held portable digital multimeter (RS Components Ltd)

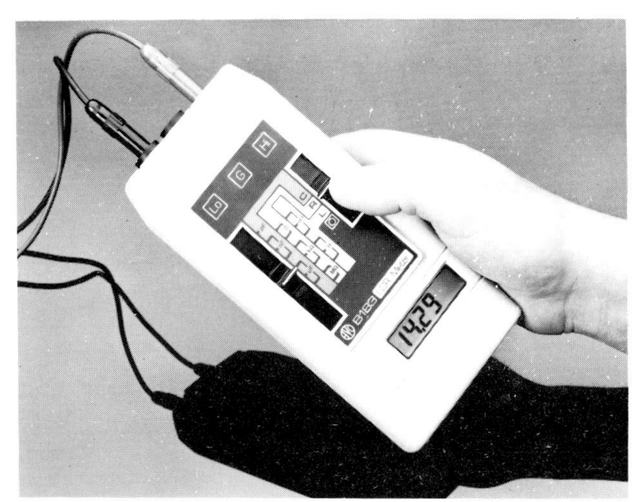

Photo 3.2 Avo B183 digital LCR meter, capable of measuring inductance, capacitance and resistance (Thorn EMI Instruments)

wave, 1.732. Digital multimeter crest factor specifications of around 7:1 and above are available.

Current may be displayed by measuring the voltage which the current creates across a known internal resistance. When measuring alternating currents the AC voltage across the resistor is applied to the RMS converter before measurement.

Resistance may be displayed by measuring the voltage developed across a known internal resistance in series with the unknown resistance, caused by a reference voltage source within the meter. By suitable scaling within the meter the resistance value may thus be displayed.

Switching between ranges when taking a measurement is usually a manual operation, but some modern digital multimeters have an auto-ranging facility, which electronically senses the level of the measurand signal and selects the appropriate range for display. Digital multimeters of the cheaper kind, which do not have auto-ranging functions, generally have an over-range indication where the most significant digit of the digital display becomes a

Photo 3.3 Philips PM 2534 digital system multimeter. The liquid crystal display shows GPIB address selection (Pye Unicam)

Photo 3.4 Fluke 70 series of digital multimeters. Although display is of the liquid crystal form the multimeters boast digital *and* analog bar graph display of information

'1' with no other digits displayed. Over-range protection is normally built-in to all digital multimeter signal conversion circuits, so that connection to large, potentially harmful voltages or currents cannot damage the instrument.

Analog-to-digital conversion

The process of converting an analog signal to one of a digital nature is fairly easy in principle. The analog signal is merely sampled, that is, a measurement of the signal value is taken at regular intervals, and each sample is presented in a binary digital way; so that, for instance, an analog voltage of 1.5 V may be presented in a digital way as, say, 1010 – a binary digital number. In the case of the common digital multimeter, this binary number is then decoded to drive the digital display, where the decimal digits '1.5' are displayed.

Under certain conditions the string of sampled values is perfectly representative of the measurand signal. The main condition which governs the conversion is that the samples should be taken at close enough intervals so that all variations, however fast, within the measurand signal are converted also. Put another way, if the analog measurand signal consists of frequency components in the range from DC to f_{max}Hz, then it must be sampled at a rate of $2f_{max}$ samples per second.

In the case of a general-purpose digital multimeter, which is used to measure a fixed, or very slowly changing, voltage or current, this is no problem and samples can be taken at very low rates. If the signal varies at a maximum frequency of, say, I Hz, then a sampling rate of twice per second is perfectly adequate to represent the measurand digitally. A digital multimeter whose display changed much more often than this would prove at best irritating and at worst unreadable anyway. Most general-purpose digital multimeters do, in fact, have a sampling rate of around 2 to 2.5 samples per second.

The analog-to-digital converter circuit used in such digital multimeters is usually a **dual-slope converter.** It is shown in block diagram form in Figure 3.2. The input of the integrator is electroni-

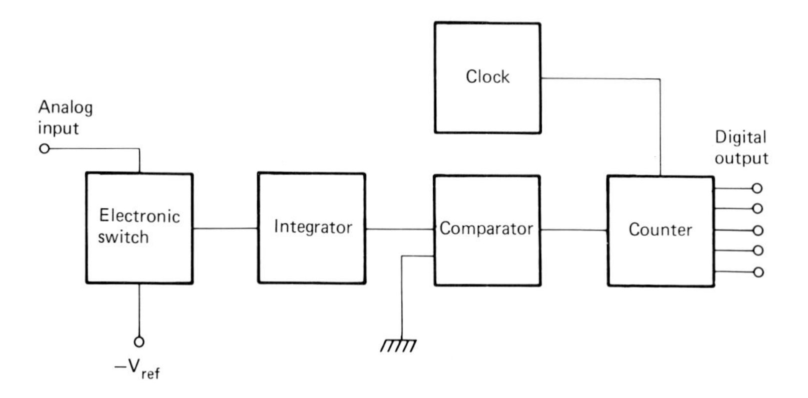

Figure 3.2 Block diagram of a dual-slope analog-to-digital converter

cally switched between a reference voltage $-V_{ref}$ and the voltage to be converted V_{in}. By integrating the analog voltage for a fixed period of time t_1, as the integrator output voltage ramps up from zero, then counting the clock pulses for the variable time t_2 which the integrator takes for its output voltage to ramp back down to zero, a very accurate analog-to-digital conversion can be made, because the time t_2 is proportional to the analog voltage.

A couple of examples should clarify converter operation. If a voltage of, say, 1.999 V, is measured, the integrator output voltage (Figure 3.3) increases in time t_1, to the full scale voltage shown. As the integrator ramps back down to zero 1999 clock pulses are counted during time t_2, and the display reads '1.999'. When the input voltage is 1 V, the integrator output voltage only reaches the half scale voltage in time t_1, so only 1000 clock pulses are counted in time t_2, before the integrator output ramps back down to zero. The display now reads '1.000 V'.

Converter circuits like the dual-slope converter are ideally suited for general-purpose digital multimeter applications: they are cheap and highly accurate. However, their slow-speed operation precludes them

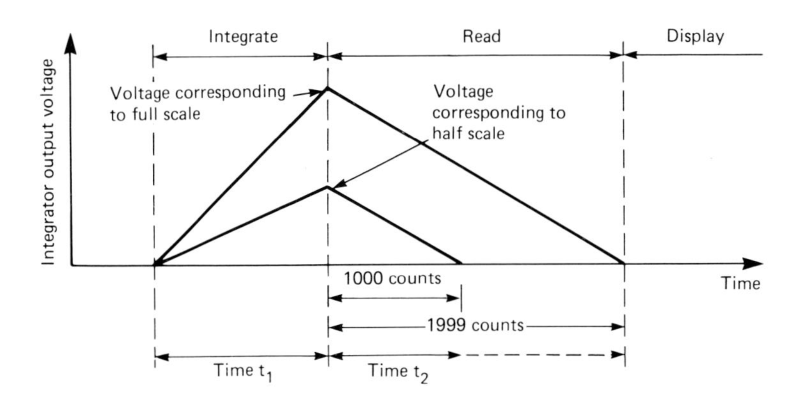

Figure 3.3 Illustrating the output voltage of the integrator in a dual-slope analog-to-digital converter, for two possible values of input voltage

from use as analog-to-digital converters in high-speed applications. Digital multimeters used in systems measurement applications, where measurements may be required hundreds of times a second, must use faster (and usually more expensive) converters. One converter typically used in such applications is the **successive approximation converter**, a block diagram of which is shown in Figure 3.4. The successive approximation converter is an example of an analog-to-digital converter which uses a digital-to-analog converter within itself (digital-to-analog converters are inherently much simpler, not relying

Figure 3.4 Block diagram of a successive approximation analog-to-digital converter

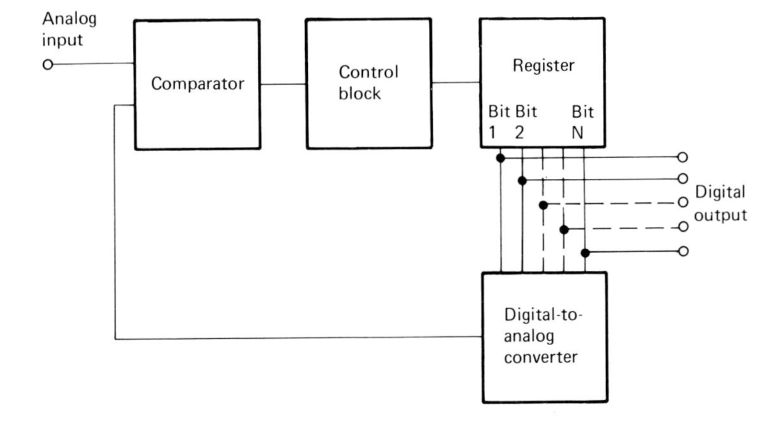

on timed samples to function) as part of normal circuit operation. In essence, the digital output of the converter is reconverted back to analog by the digital-to-analog converter, then compared with the analog input voltage by a comparator. The comparator output is logic 1 if the analog input signal is greater than the digital-to-analog converted signal, but logic 0 if the converted signal is larger than the analog signal. At commencement of operation, the digital output is assumed to be zero, therefore the output of the digital-to-analog converter is also zero. Thus, any applied analog input signal causes the comparator output to be logic 1. The first clock pulse causes the control block to set the most significant bit, bit 1, of the register to 1, and so the digital-to-analog converter output increases. If the applied input voltage is still higher than this, the output of the comparator remains at 1 and the next clock pulse causes the control block to set the next most significant bit, bit 2, to 1, and so on.

When the digital-to-analog converter output voltage becomes higher than the applied voltage, the comparator output changes to logic 0 and the next clock pulse causes the control block to reset the last bit to 0 before setting the next bit to 1. Thus the process can be seen to be a number of successive approximations of the applied voltage (hence the name). Figure 3.5 shows a timing diagram of the analog output voltage of the digital-to-analog converter within an 8-bit successive approximation analog-to-digital converter. After the final approximation has taken place, the converter's digital output represents an accurate conversion of the applied analog input, which may be directly displayed.

The total conversion time of a successive approximation converter is equal to N clock cycles, where N is the number of bits. The conversion time is thus independent of the input voltage (unlike the dual-

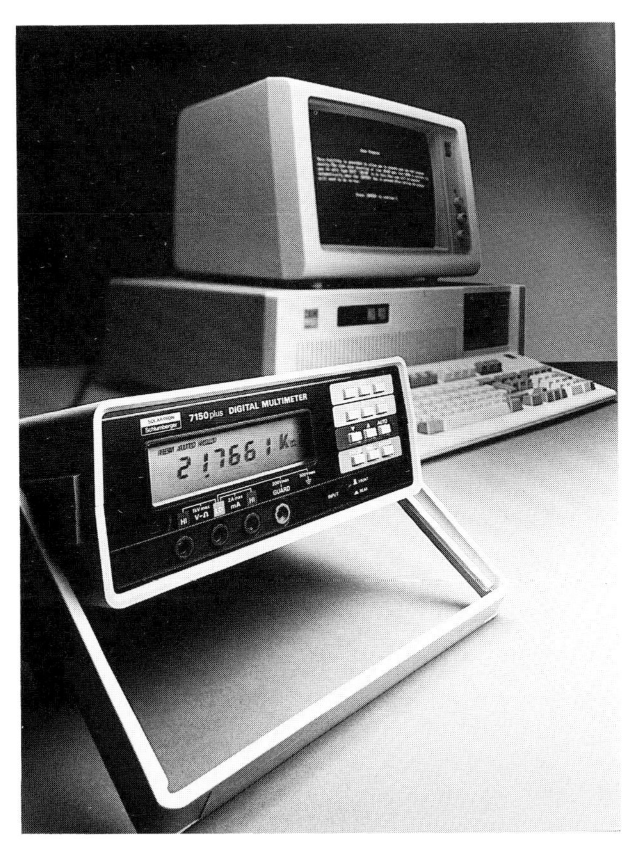

Photo 3.5 Solartron 7150plus digital multimeter (Schlumberger)

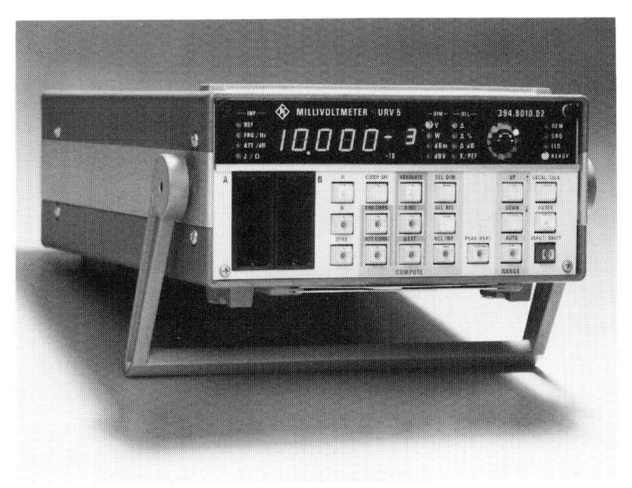

Photo 3.6 Rohde & Schwarz digital millivoltmeter (Rohde & Schwarz)

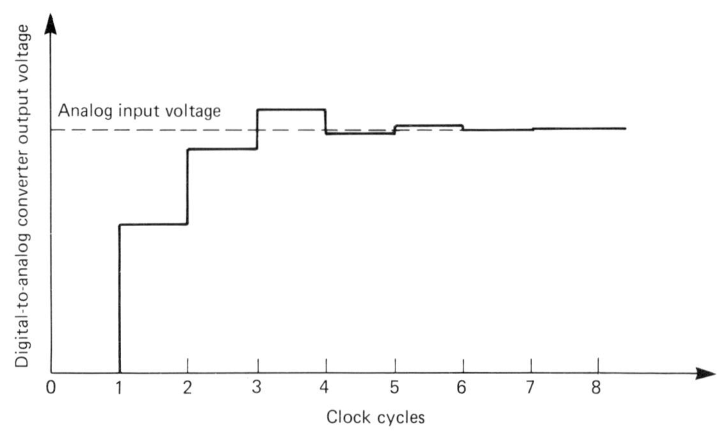

Figure 3.5
Possible timing diagram of the analog output voltage of the digital-to-analog converter in a successive approximation analog-to-digital converter. The voltage gets closer to that of the input voltage with each approximation

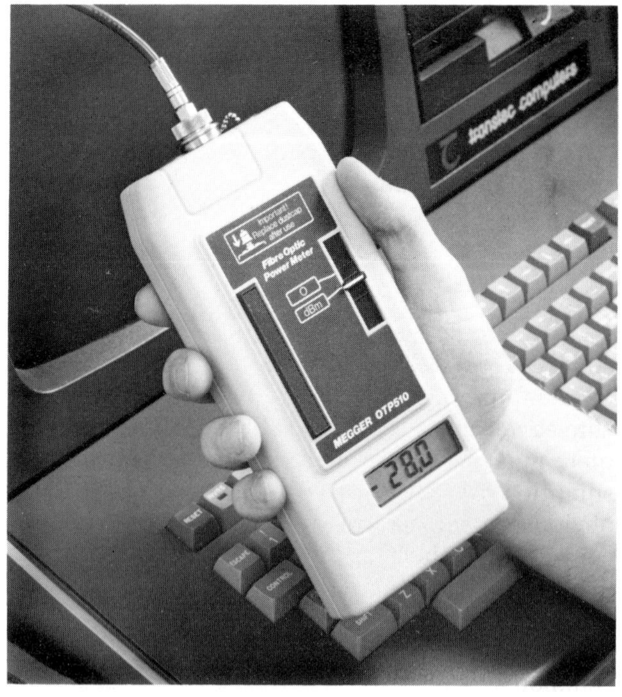

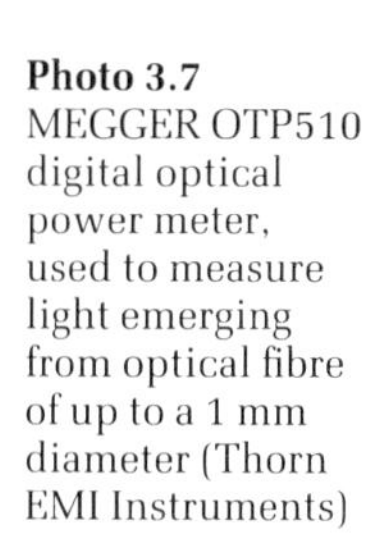
Photo 3.7
MEGGER OTP510 digital optical power meter, used to measure light emerging from optical fibre of up to a 1 mm diameter (Thorn EMI Instruments)

slope converter's conversion time) and, in fact, is only limited by circuit delays. Many more samples per second can thus be taken using this technique, and, typically, digital multimeters which take 200

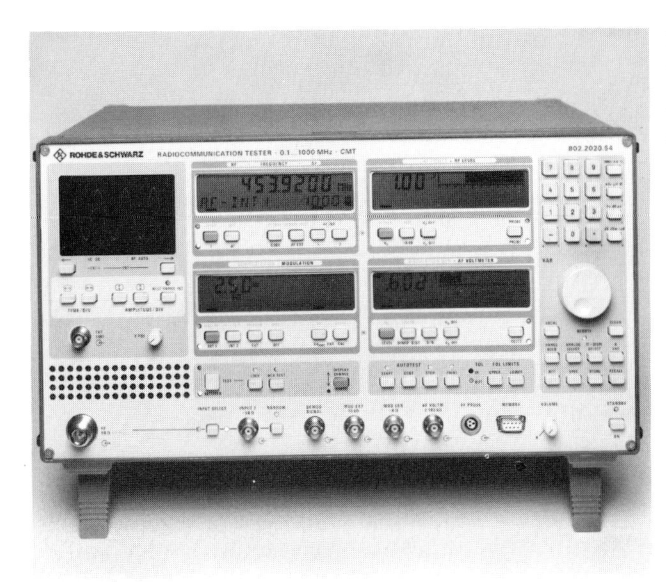

Photo 3.8 Rohde & Schwarz CMT radiocommunication tester, using a number of digital displays (Rohde & Schwarz)

to 1000 samples per second are available, depending on the number of bits converted.

Displays

Most general-purpose digital multimeters use seven-segment liquid crystal displays. Low power consumption of LCDs means that, with low current consumption signal conversion and analog-to-digital conversion circuits (say, of CMOS origin), the complete multimeter may have an overall current consumption of only a few microamps. This is an obvious advantage in portable, battery-powered equipment. Bench test equipment, however, being mains powered, does not have this power consumption limitation and light-emitting diode displays or, sometimes, vacuum fluorescent displays are common in larger digital multimeters used in systems applications.

Generally between four and eight digits are displayed by a digital multimeter. A decimal point is usual, which moves along the display corresponding to the range being displayed. Usually the most

Photo 3.9 Thandar TM357 and TM358 digital multimeters (Thurlby-Thandar)

significant digit in the display is not of seven segments, but merely displays the number I or nothing at all. For this reason, a meter with three full seven-segment digits plus a most significant digit of this type is said to have a **three-and-a-half digit display,** and most hand-held, portable digital multimeters of a general-purpose nature have displays of this size. Maximum indication possible using a 3½-digit display can only be '1.999', but more expensive digital multimeters are available, where greater resolution is required, with up to 8½-digit displays.

Advantages of digital meters

In Chapter 2 it was noted that the accuracy of analog meters is determined primarily by the moving-coil meter movement itself, together with the position of the pointer on the scale. Thus the highest accuracy a moving-coil meter can have is given by its percentage accuracy at full scale deflection. Other sources of error are loading of the measurand circuit by the input resistance of the meter when measuring voltage, and the resolution with which the reading is taken by an observer.

In a digital meter these three error sources are eliminated. The display is totally electronic, with no moving parts, so cannot possibly be inaccurate – whatever the preceding circuit supplies it with, the display presents as a numerical read-out. This numerical read-out is not subject to resolution problems such as parallax or pointer position errors. Finally, the electronic circuits used in a digital meter present an extremely high input resistance (typically 10MΩ) to the measurand circuit, thus preventing the possibility of loading errors in all but the highest resistance measurand circuits. Instead accuracy and resolution of a digital meter are dependent on the accuracy and resolution of the internal electronic circuits, so that errors in indicated measurements are due mainly to component inaccuracies and poor calibration, although sometimes poor design may be a culprit.

Typical accuracy of a low-cost, general-purpose multimeter is less than ± 1 per cent of the reading (whatever the reading – not the full scale deflection limitation of analog meters), and resolution is about one in 2000 parts. Accuracy of high-quality digital multimeters is generally less than ± 0.005 per cent of reading with a resolution of about one part in 100 000.

Overall, the digital multimeter, and digital meters in general, have many advantages over analog multimeters and meters. The need for greater accuracy in manufacturing and servicing has meant they have become increasingly popular as test equipment instruments. Although general-purpose digital multimeters tend to be slightly more expensive than analog counterparts, analog meters of the accuracy and resolution afforded by high-quality digital meters could never be made, whatever the price. The digital meter has created its own marketplace, which the analog meter can never overturn.

Photo 3.10 Thurlby 1905a intelligent digital multimeter (Thurlby-Thandar)

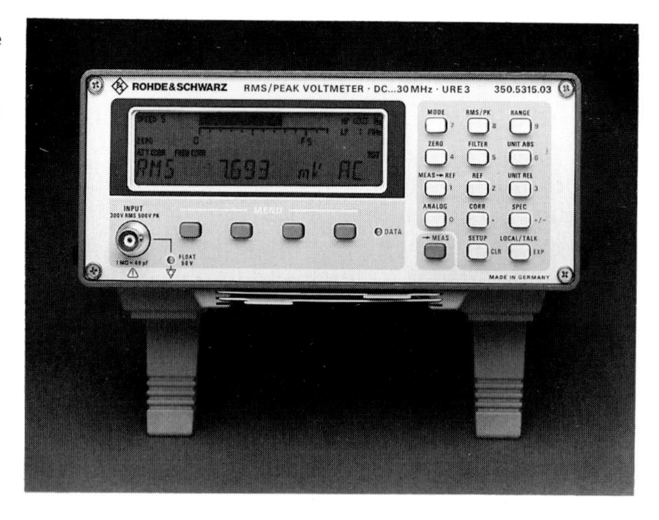

Photo 3.11 Rohde & Schwarz URE3 RMS/peak voltmeter (Rohde & Schwarz)

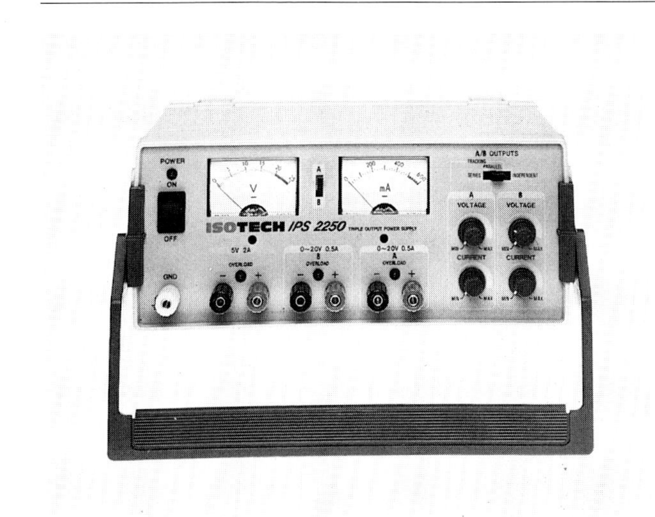

Photo 3.12 A bench power supply (RS Components Ltd)

4
Oscilloscopes

Strictly speaking, an oscilloscope is *any* device that can display waveforms. The term 'oscilloscope' is derived from the Latin word *oscillare*, meaning to swing backwards and forwards, and the Greek *skopein*, meaning to observe, aim at, examine. A number of electronic test instruments have this capability: XY plotters and pen recorders are two examples. However, the term has come to be used in a general way to refer to the particular type of test instrument covered in this chapter. Some other test instruments which display waveforms and thus qualify for inclusion into the category are covered in other chapters.

After all kinds of analog and digital meters the oscilloscope is the next most common laboratory test instrument. It is by far, though, the *most* useful test instrument, allowing the innards of a circuit to be analysed in *real-time*; i.e., what the oscilloscope displays is actually occurring in the circuit. I'm sure my old university lecturer wouldn't object if I bring to mind his belief that he knew when a student was getting the hang of electronics as a subject when the first test instrument used to study a circuit was the oscilloscope. What he meant by this, of course, is that the oscilloscope allows the user to 'see' the electronic action of a circuit, thereby confirming correct (or incorrect) operation – students who didn't use it didn't really know if the circuit was working correctly. The magic moment when the oscilloscope was the first instrument a student fetched from stores represented a milestone which he personally took to mean that he was doing his job well and getting somewhere.

Compared with a meter, say, an analog moving-coil meter, which can only display direct voltages and currents (or, at most, very slowly varying ones), the oscilloscope can display alternating voltages and currents. In effect, a meter can only display a measurement in one dimension, that is, the measurand's amplitude at any given time. The oscilloscope, on the other hand, displays two dimensions, the measurand's amplitude against the second dimension of time. In this way rapidly varying amplitudes may be easily displayed.

This ability of the oscilloscope to display two-dimensional measurements arises from the type of display device it uses. The more technical description of what is normally called an oscilloscope is a 'cathode ray oscilloscope' (CRO), which gives a clue to the display used: a cathode ray tube (CRT). Operation and composition of cathode ray tubes are covered in detail in Appendix 1; all we need to know here is that the cathode ray tube may be used to display on its screen the two dimensions of amplitude and time. This is done by moving the electron beam in the cathode ray tube across the screen in a controlled way, so that a representation of the measurand is displayed, the vertical amplitude of which corresponds to the measurand's amplitude and the horizontal amplitude of which corresponds to units of time over which the measurand is observed.

What the oscilloscope displays

The real-time or general-purpose oscilloscope is essentially used for displaying a representation of periodically varying (that is, repetitive) measurands. Such a measurand is shown in Figure 4.1a, and is seen to be a fairly simple periodic voltage waveform. To display the waveform the oscilloscope first has to break it down into manageable, screen-sized portions, as in Figure 4.1b. These portions of the

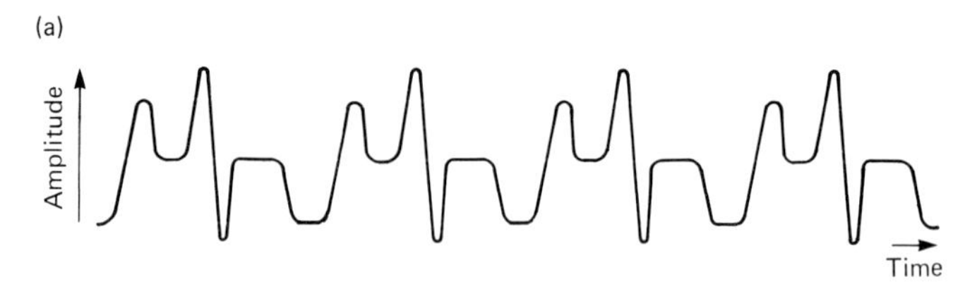

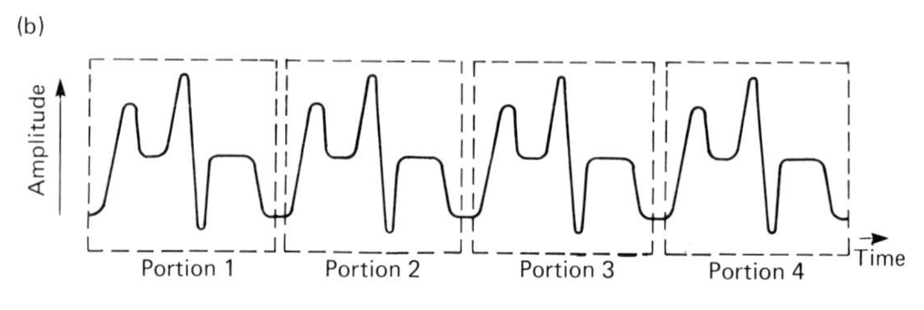

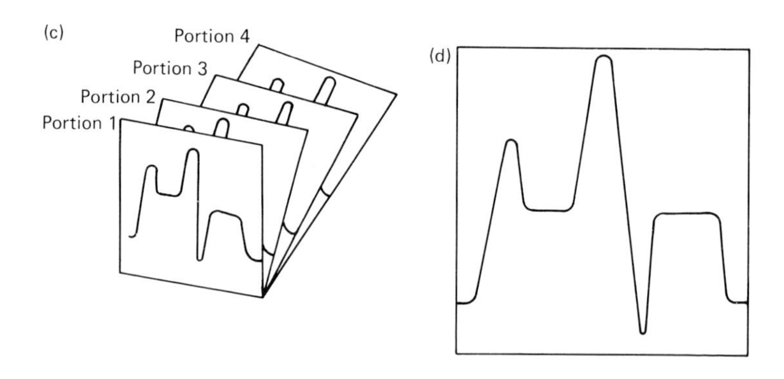

Figure 4.1 Possible periodically varying measurand (a) as it occurs (b) broken down by the oscilloscope into portions (c) displayed in turn on the oscilloscope screen (d) as it is viewed by the observer

waveform are then displayed in turn on the screen, as in Figure 4.1c, just like leafing through a child's flick-book – the only difference being that the pictures in a flick-book are minutely different so that flicking through them creates the illusion of movement, whereas the oscilloscope's pictures are all identical, creating the illusion of a constant picture. So, with a continuous periodic waveform, the apparent displayed representation would be just one of these portions, as in Figure 4.1d. The final display is, in fact, a graph of voltage against time for this apparently single portion. This representation of a

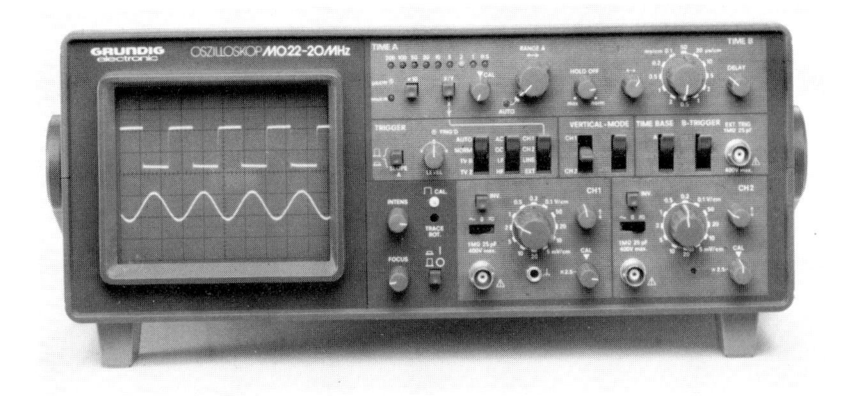

Photo 4.1 Grundig M022 oscilloscope (Electronic Brokers)

measurand which an oscilloscope makes on its screen is commonly called a **trace.** The term refers to the way the electron beam seems to trace out on the screen the displayed waveform.

It's important to remember, though, that the trace on the screen is made up from a continuously changing number of separate cycles. The effect is similar to that observed when watching cinema films, which are a rapidly changing number of 'stills', or a television picture, which is made up from many rapidly changing 'frames'. It's the persistence of the human eye which 'joins up' the separate images of each displayed cycle, making the brain conclude that a single, steady, picture is there.

The basic oscilloscope

More complex oscilloscopes can display non-periodic waveforms, too. We'll discuss those later but, for the time being, our description of a device capable of displaying periodic waveforms is perfectly adequate and provides a good model for us to consider. We can now look at the various parts of the oscilloscope circuit which allow it to function as we've seen.

Figure 4.2 shows the action of a basic oscilloscope

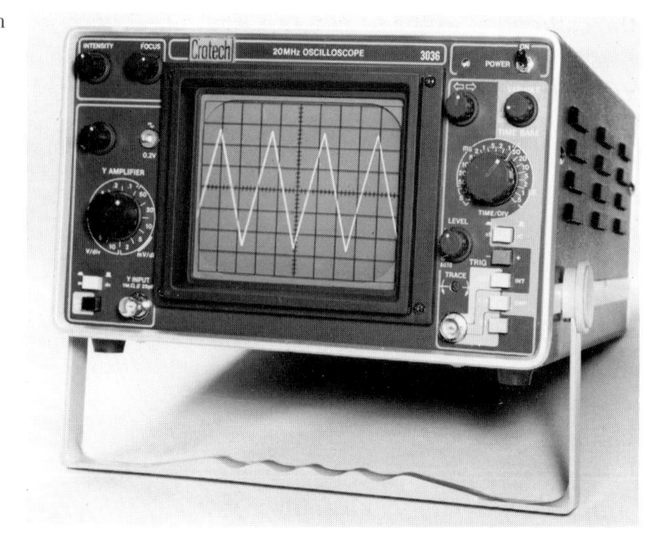

Photo 4.2 Crotech 3036 single trace oscilloscope (Crotech Instruments)

in block diagram form. Waveforms at various points around the diagram illustrate circuit action.

To enable the image of the measurand to be displayed, the electron beam must be moved across

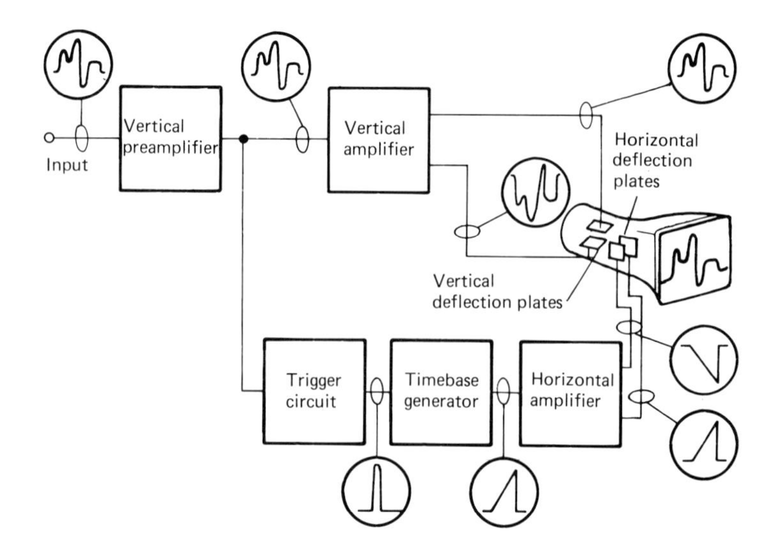

Figure 4.2 Block diagram of a basic oscilloscope with waveforms at various points

the cathode ray tube screen in a controlled manner. First, it has to move horizontally from left to right (as viewed). When the beam reaches the right-hand side of the screen it must return to the left to begin a new trace. This whole procedure must occur at regular intervals. Second, the amplitude of the displayed waveform must correspond to the amplitude of the measurand, so the electron beam must be varied in a vertical manner, too. Horizontal movement of the beam creates an axis corresponding to time, vertical movement creates an axis corresponding to amplitude.

Horizontal beam movement is effected by applying voltages, called **sweep** or **timebase** voltages, to the horizontal deflection plates of the cathode ray tube. A possible time-base voltage waveform is shown in Figure 4.3 – it's basically a ramp voltage. This could be the voltage applied to one of the horizontal deflection plates: a second time-base voltage, the exact inverse of this, would be applied to the other plate. When the voltage is low, say at *point 1*, the beam is deflected to the left-hand side of the screen. When the voltage is at *point 2* the beam is not deflected and so is in the middle of the screen. Finally, when the voltage is at *point 3* the beam is deflected to the right-hand side of the screen, and so on. The shape of the timebase voltage tells us how the beam traces across the screen. When it moves from left to right it does so in a steady manner, at a constant rate determined by the steepness of the increasing voltage. On its return trip from right to left, however (known as the **flyback**), it instantly jumps back ready to start the slower left to right

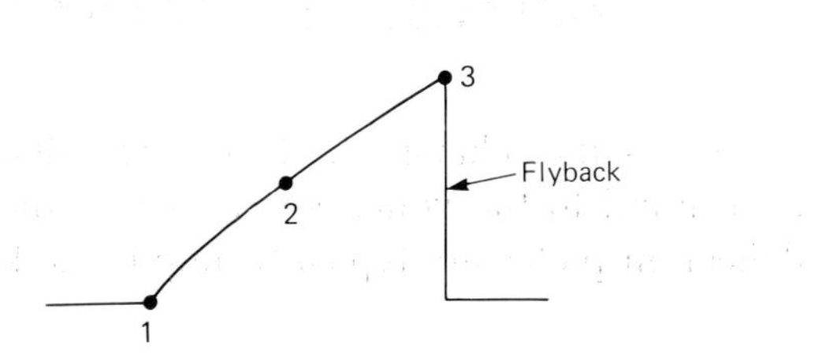

Figure 4.3 A possible timebase voltage waveform. At point 1 the beam is deflected to the left-hand side of the screen (as viewed); at point 2 the beam is undeflected, and lies at the screen centre; at point 3 the beam is deflected to the right-hand side of the screen. Finally, the beam flies back to start again from point 1

movement again. An instant jump back from right to left, that is, in zero time, ensures that a user does not see a trace on the screen as the electron beam flies back. In reality, timebase generator circuits cannot produce such an ideal timebase voltage, and typically their ramps do not increase at a constant rate but vary slightly, and their flybacks take a finite but very small time.

The length of the ramp, that is, the time the beam takes to sweep from left to right on the screen, is user-adjustable to allow different measurands with different timeslots to be displayed.

In order that the same part of the waveform is displayed by each sweep of the beam across the screen, a **trigger** circuit is used to cause a pulse to occur, which literally kick-starts the timebase generator, as the input signal reaches a certain voltage. This circuit is user-adjustable so that the point in the timeslot at which the sweep starts can be selected.

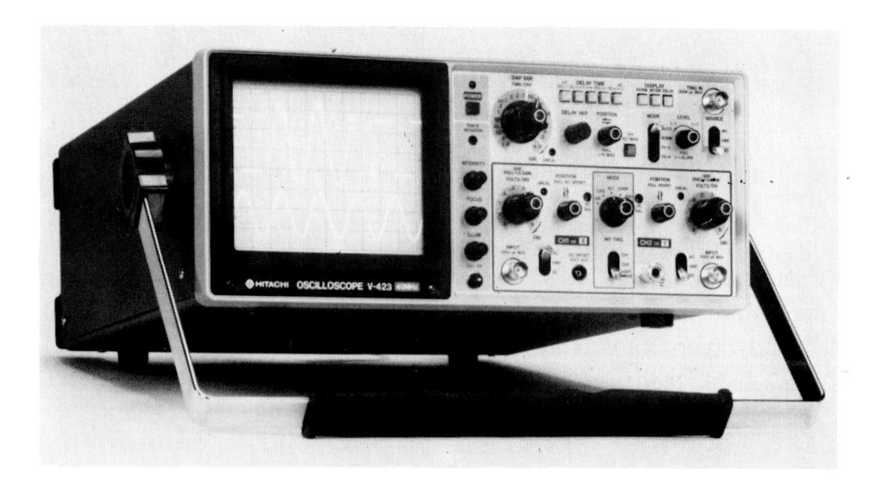

Photo 4.3 Hitachi V-423 40 MHz oscilloscope (Thurlby Electronics)

Circuits that change the incoming voltage of the measurand to the voltages required by the vertical deflection plates are typically amplifiers, known as

vertical amplifiers. Two stages of amplification are used. The first stage amplifier, known as the **vertical preamplifier** converts the incoming signal waveform to a standard-sized waveform. To cater for different amplitudes of input signal, the gain of the vertical preamplifier must therefore be user-controlled, although the gain of the vertical amplifier itself is fixed.

There are a number of additions to this basic circuit which will make the oscilloscope far more useful. The additions are shown in a more complex block diagram in Figure 4.4. At present, it is only capable of displaying DC waveforms, so the first addition is of a capacitor and switch at the Y input. This allows the user to select between AC or DC waveforms.

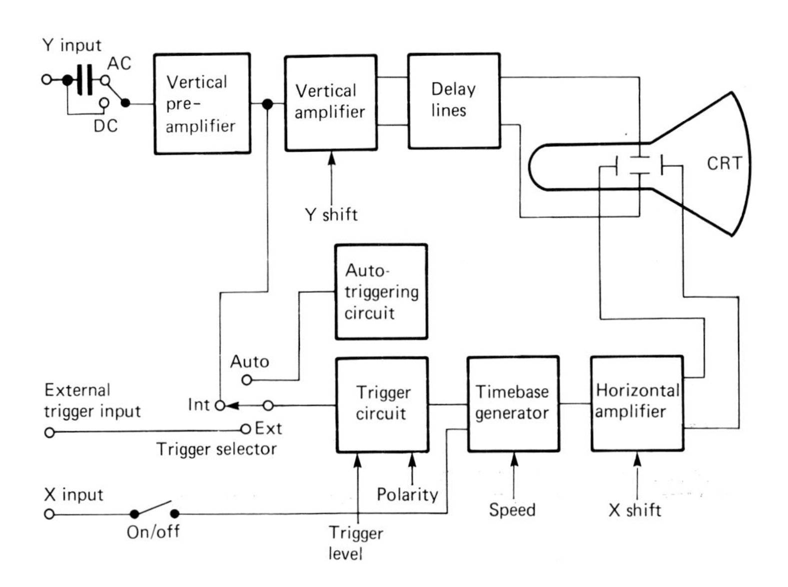

Figure 4.4 More complex block diagram of a basic oscilloscope, with typical additions

If large DC voltages, or AC voltages with a large DC bias, are to be displayed, a user-adjustable control to adjust vertical deflection voltages must be included. This is the **Y shift** control, which allows zero Y voltage to correspond with, say, the centre horizontal line. A similar control adjusts the horizontal deflection voltages, known as the **X shift**. The X shift allows the complete timeslot of the observed waveform to be moved left or right over the screen.

In many instances the user may require the oscilloscope to be triggered not from the viewed waveform, but from an external source. Thus the viewed waveform can be observed in direct comparison with a separately occurring trigger source. The **trigger selector** provides this function. The point at which triggering of the timebase generator occurs is adjusted by the **trigger level** control. The addition of an **auto-triggering circuit** allows automatic triggering of the timebase generator, without the need to adjust trigger level. Triggering may take place on a positive- or negative-going edge, depending on the **trigger polarity** switch.

One final external input is the **X input**, which allows the timebase generator to be disabled and an external voltage to control horizontal deflection.

Photo 4.4 Philips PM 3295 350 MHz oscilloscope (Pye Unicam)

What is described here, although it is a real-time oscilloscope, is not particularly common. Most practical oscilloscopes have a **dual-trace display**; that is, two traces are available so that two different waveforms can be displayed. In dual-trace oscilloscopes the electron beam is switched rapidly from one waveform to the other so that the eye perceives a constant display of two waveforms. Oscilloscopes with four or more traces are also available., Of course, sharing the single electron beam between two, four, or more traces means that less and less time is spent by the beam displaying any one waveform and the traces will be correspondingly dimmer. Some oscilloscopes have separate electron guns to defeat this problem, and are known as **dual-beam oscilloscopes.**

Advanced, but still real-time, oscilloscopes exist which have other features, too. Separate timebase generators (one for each trace), timebase generator delay facilities (to allow displayed traces to be offset), and input waveform delay, facilities (to counteract the effect of circuit delays which would otherwise render impossible the observation of rapidly occurring signal changes), are some of the many features.

However complex real-time oscilloscopes are, they can only be used to observe rapidly repetitive waveforms. Also available are **non-real-time oscilloscopes** which allow the observation of non-repetitive, that is, singly occurring waveforms, which real-time oscilloscopes could not display. An example of such a singly occurring waveform is the interrupt request signal to a microprocessor. This may only take place once every so often and is of only a few microseconds duration. To display this kind of waveform a **storage oscilloscope** can be used, in which the displayed trace represents a *single* timeslot of the complete waveform that is stored within the oscilloscope.

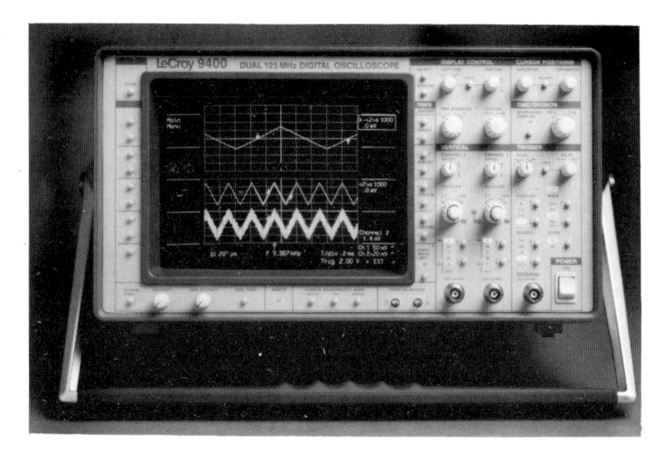

Photo 4.5 LeCroy 9400 125 MHz digital storage oscilloscope (LeCroy Resarch Systems)

There are two main types of storage oscilloscope: one which depends on a special cathode ray tube which maintains its display long after the electron beam has swept across it (see Figure A1.3, in Appendix 1); and another which samples and stores the waveform digitally, ready for later recall and display. The **cathode ray tube storage oscilloscope** is cheaper than the **digital storage oscilloscope** (DSO), but is correspondingly less versatile. The trace stored within a storage cathode ray tube cannot be repositioned or altered, whereas the digitally stored trace can be moved, magnified, contracted, erased, and displayed yet again at the user's whim, long after the recorded event. Most storage oscilloscopes have the facility to be switched between real-time (that is, non-storage) and non-real-time (that is, storage) operation, so two separate oscilloscopes are not required to give all functions.

The system of sampling the input waveform used by the digital storage oscilloscope is also used in a type of oscilloscope called a **sampling oscilloscope.** The sampling oscilloscope, however, does not store the samples it takes of the waveform, merely holding each sample for one sampling period only, when it is replaced by the next sample. Extremely large bandwidths are possible using this technique (over 10 GHz) and it is used mainly in specialised equipment. Figure 4.5 gives a family tree of various types of available oscilloscope.

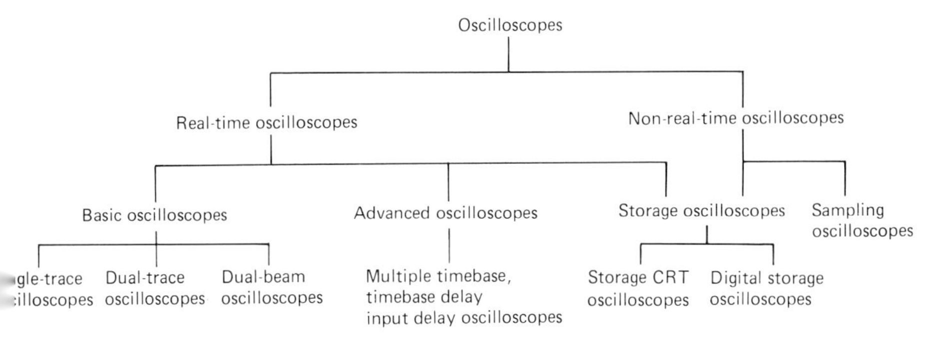

Figure 4.5 Family tree of the main types of oscilloscope

Specifications

We now turn our attention to those aspects of oscilloscopes which define one device as being better or worse than another. Put another way, what should a user be looking for in an oscilloscope?

First, bandwidth. Like any electronic circuit, the oscilloscope only passes a limited range of frequency components. Components outside the bandwidth range are drastically reduced in amplitude. Generally, the bandwidth of a circuit is defined as the band of frequencies the circuit passes, over which the circuit's power amplification falls within a specified fraction (usually one half) of the maximum.

In all circuits this means the output signal consists of frequency components, of the correct power, within the bandwidth and frequency components, with reduced power outside the bandwidth. In the case of the oscilloscope the output is the displayed waveform, and is not one of power but of amplitude. Half power corresponds to a reduction of component amplitude of $\sqrt{3}$; that is, about 0.7 of the mid-band component amplitude. So the user can still see frequency components outside the oscilloscope's bandwidth, but they are of reduced size.

Photo 4.6 LeCroy Waveform-catalyst digital storage oscilloscope, shown in use with personal computer (LeCroy Research Systems)

Obviously, the wider the bandwidth, the better the oscilloscope. Bandwidths of 10 MHz or 20 MHz are common in general-purpose oscilloscopes, but oscilloscopes with bandwidths up to around 250 MHz are available for higher frequency work. More specialised oscilloscopes (very expensive) have bandwidths up to 1 GHz.

Also of consequence and related to the bandwidth is the range of amplification factors of the vertical amplifier. The greater the amplification the smaller the waveform that can be displayed on the screen. Oscilloscope screen displays are marked in a grid of 10 by 8 centimetre square divisions, known as the **graticule.** Amplification factors are denoted in volts/division or volts/centimetre, so that a factor such as 10 volts/div corresponds to a total screen height of 80 volts. Typical amplification factor ranges are from about 10 millivolts/div to 5 volts/div. The better the oscilloscope the greater this range, so higher quality oscilloscopes have ranges which extend down to 1 millivolt/div and up to 50 volts/div.

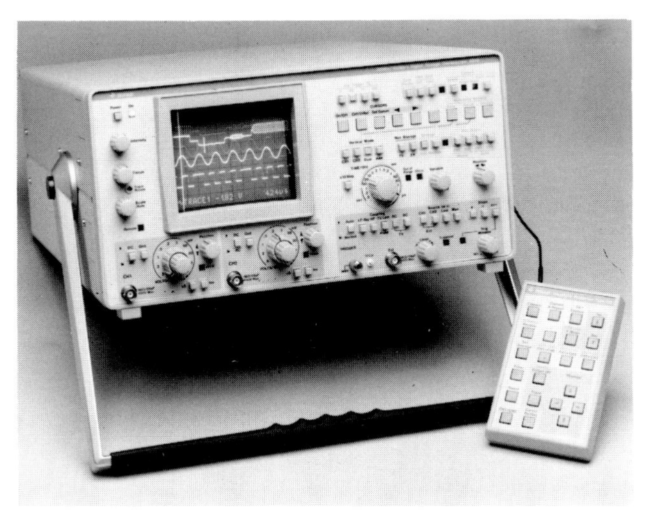

Photo 4.7 Gould 4050 digital storage oscilloscope (Gould Electronics)

Bandwidth considerations discussed here apply to the Y, i.e. vertical, circuitry within the oscilloscope. However, there's little point in having a high Y bandwidth if the X circuitry doesn't have a similar high frequency ability. The X 'bandwidth' corresponds to the range of settings available in the oscilloscope's timebase generator controls. Timebases in the range of 2 seconds to about 10 microseconds (times given indicate the time taken for the beam to sweep from the left-hand to the right-hand side of the CRT screen – times are generally denoted per division, e.g. 0.2 second/div to 1 microsecond/div) are common, but with oscilloscopes of higher Y bandwidths, total timebase times must decrease to about 10 nanoseconds (i.e. 1 nanosecond/div) for Y bandwidths of around 250 MHz, or less for specialised oscilloscopes.

Oscilloscope accessories

The standard input resistance of oscilloscope vertical amplifiers is 1 MΩ. Such a high resistance would, you might think, allow measurements of most measurands to be taken without the oscilloscope loading the measurand circuit in any appreciable way. This should be reflected in the accuracy of the measurement. However, such a high input resistance necessitates the use of a screened input lead to connect the oscilloscope to the measurand circuit, to prevent excessive hum pick-up. Typical screened lead capacitances are around 50 to 100 picofared/metre (one metre of input lead is a convenient length) and this, coupled with the vertical amplifiers' own input capacitances of around 15 to 50 pF (depending on make and model), makes a total input capacitance of about 100 to 150 pF. With input capacitances of this order, giving a reactance of around 140 Ω at 10 MHz, you can see that considerable loading of high frequency measurands could easily occur, thus affecting measurement accuracy.

The usual solution to this problem is to use a standard accessory: the **passive divider probe**. Such probes have an attenuator within, which has the effect of increasing the resistance presented to the measurand circuit, while decreasing the capacitance. A typical probe attenuator is 10:1, which presents a resistance of ten times that of the vertical input of the oscilloscope, i.e. 10 MΩ, and one-tenth the input capacitance, i.e. about 10 to 15 pF, effectively reducing loading. The attenuation which a passive 10 × divider probe causes can be counteracted merely by increasing the vertical amplifier's gain by the same amount.

In addition to the common 10 x probe, passive probes of 100 × and even 1000 × attenuation are available. These would only be usable, however, to allow measurements of relatively large measurands, as the oscilloscope's vertical amplifier may not have sufficient gain to counteract the higher attenuation.

Active probes are an alternative, employing a field effect transistor amplifier which presents a high

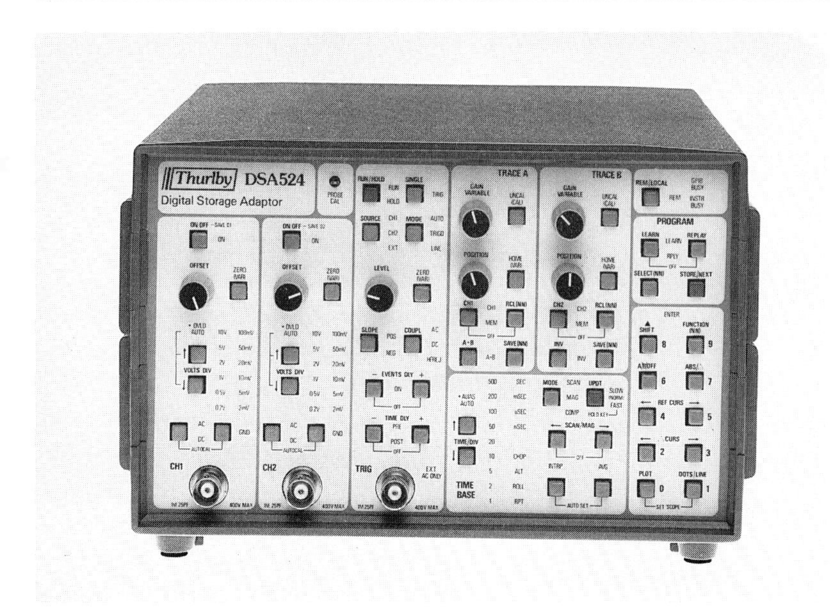

Photo 4.8 Thurlby DSA524 digital storage adaptor (Thurlby-Thandar)

resistance and low capacitance to the measurand circuit, with unity gain (that is, no attenuation) or even amplification. Active probes are more expensive than their passive counterparts and require a power supply, generally a small cell, for the internal field effect transistor amplifier.

The oscilloscope is a voltage measuring device, of course, but that doesn't prevent it from being adapted to measure other electrical parameters, such as current. The obvious way to do this is to route the measurand current through a known value resistor, then observe the voltage across the resistor. The displayed voltage corresponds exactly to the current. A more preferable method is to use a **current probe**, which uses an internal transformer to develop an equivalent voltage which is then presented to the oscilloscope. Such probes' low frequency responses are limited, but **active current probes** are available which use the Hall effect principle and allow oscilloscope current measurements down to DC.

Equipment accuracy

There are three main sources of error when using an oscilloscope to take measurements. First, although probes reduce loading of the measurand circuit to a defined amount, loading still occurs and so voltage measurements are not necessarily as accurate as the user thinks. Also concerning passive probes, the probe's internal compensation capacitor must be adjusted to keep waveform distortion and amplitude error to a minimum. As a general rule this adjustment should be done every time the oscilloscope is used. The check is quite simple: the most convenient way is to observe a 1 kHz squarewave and adjust the capacitor for clean and square transition corners. Figure 4.6a shows the observed squarewave when using a correctly compensated probe. Figure 4.6b shows the observed waveform when the probe is undercompensated and Figure 4.6c shows it when the probe is overcompensated. Many oscilloscopes have a squarewave output available from the front panel, specifically for the purpose of compensating passive probes.

The second main source of errors comes from the internal oscilloscope circuits. Errors can exist due to incorrect gain in the vertical amplifiers, or to incorrect timebase times in timebase generator

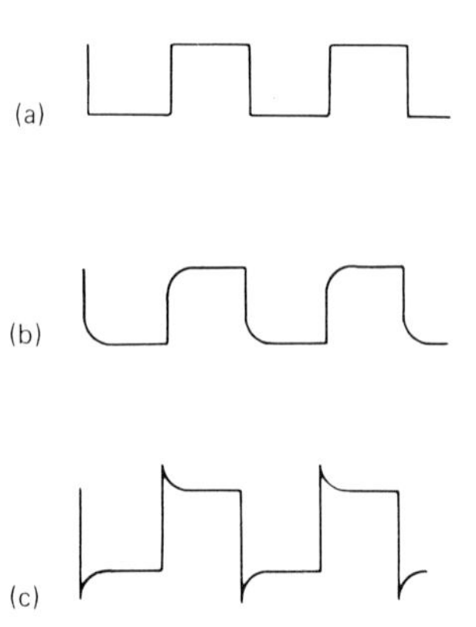

Figure 4.6 Viewed waveforms for an applied squarewave when the passive probe is adjusted (a) correctly (b) under-compensated (c) over-compensated

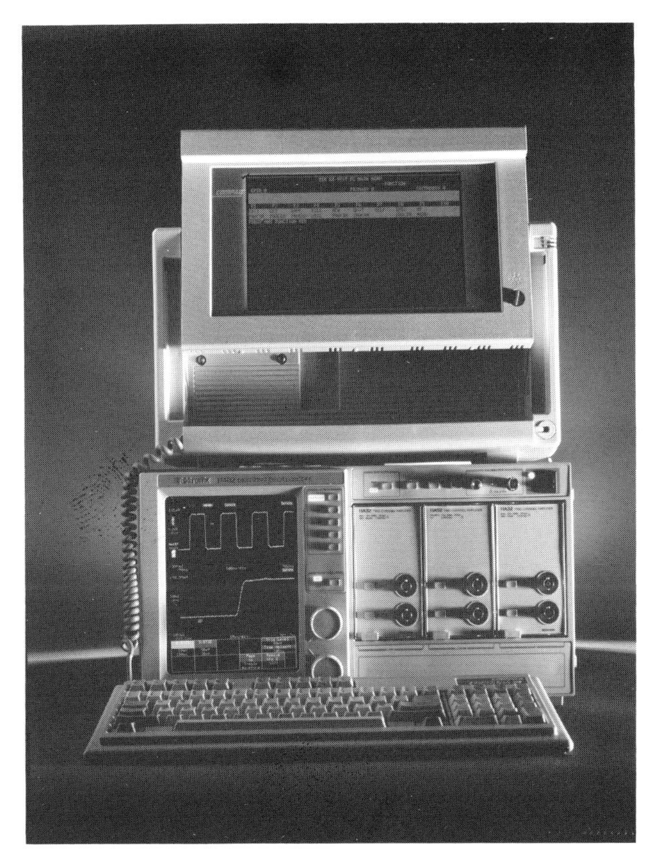

Photo 4.9 Tektronix 11402 digitising oscilloscope (Tektronix)

circuits. With the vertical amplifier controls set in the *calibration* position, an input of, say, 1 volt must produce a display corresponding to 1 volt vertical displacement on the screen. Similarly, with the timebase generator controls set in their calibration position, a pulse input of, say, 1 second duration must produce a waveform display corresponding to that horizontal displacement. The same squarewave output used to compensate probes can often be used to check and adjust for vertical amplifier and timebase generator errors.

Finally, user errors make up the third main source. There are a number of possible problems. Parallax

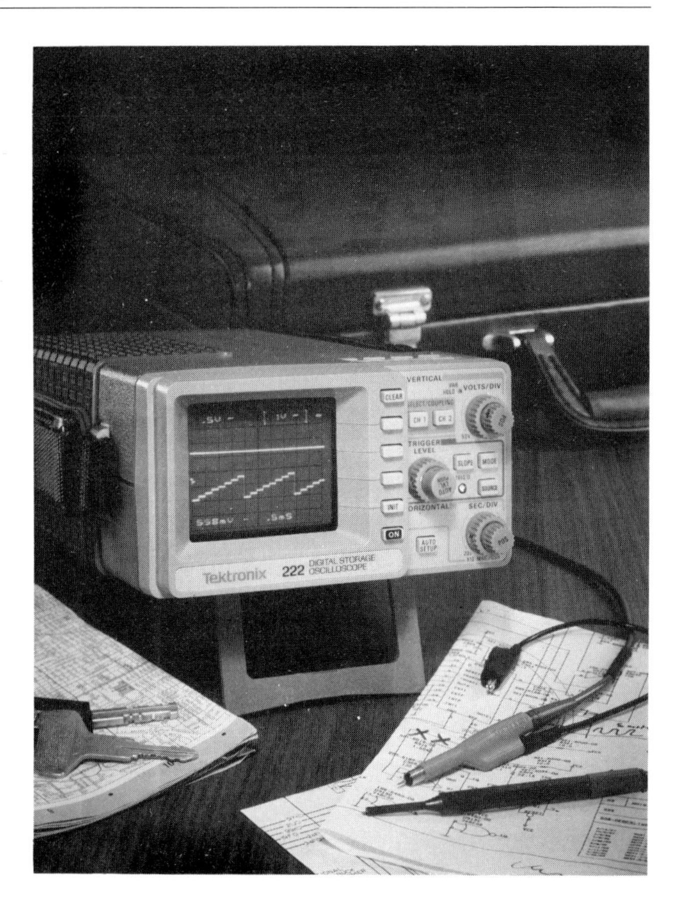

Photo 4.10
Tektronix 222 digital storage oscilloscope (Tektronix)

is the most prevalent. In most oscilloscopes the graticule is in a different plane from the layer of phosphor on which the trace is inscribed by the electron beam. So, if the user observes the screen at anything other than perpendicular, parallax errors will occur in the readings. Some oscilloscopes have graticules on the inside of the screen, virtually in the same plane as the phosphor layer, thus reducing parallax errors to a minimum.

Non-linearity can occur on the edges of cathode ray tube screens and so important measurements should be taken in the central, say 6 by 4 division, portion of the 10 by 8 division graticule, whenever possible.

As a general rule the beam intensity should be kept as low as practicable, because a dim beam produces a narrower trace than a bright beam and so it is easier to assess correctly the beam's true centre. In a similar vein, beam focus should be regularly checked throughout measurements.

New developments

The most important developments under way in oscilloscope technology are those which aim to connect the equipment via a microprocessor-controlled interface bus. This principle is discussed in another chapter, but is at least worthy of note here.

Some recent oscilloscopes have been made in a modular fashion. Such **mainframe** oscilloscopes utilize plug-in units which the user selects for personal requirements. These oscilloscopes are generally of extremely high quality, but the plug-in modular nature keeps equipment cost down to a respectable level, as the user only buys those items needed at any one time. Further purchases of other modules can then be made at later times, as needed.

Although the cathode ray tube has always previously been the only device suitable for display purposes in the oscilloscope, the liquid crystal display (LCD) has made its entry. Although liquid crystal displays have a strictly limited response, which defines their use only in low bandwidth oscilloscopes, they do have the advantage of low power consumption. Truly portable (that is, battery powered) oscilloscopes are now possible, and at least one manufacturer markets such an LCD-based oscilloscope.

5 Signal sources

More by default than by design, signal sources are generally categorised into two areas: audio frequency and radio frequency sources. Again by default, those sources which produce audio frequency signals are usually called **low frequency oscillators,** and those which produce radio frequency signals are usually called **signal generators.** However, there's no logical reason for all this and, as we'll see, a number of *illogicalities* arise with the annotation.

In short, signal sources are generally used to produce signals which are applied to circuits; to test those circuits' performances in the design, manufacture and service stages of their lives. It follows that the signal source used to test any particular circuit must be of the right type: for example, an audio amplifier could not be tested with a radio frequency signal – rather obvious, but it needs stating. This, of course, is where the main

Photo 5.1 Philips PM 5193 programmable frequency synthesiser/function generator (PYE Unicam)

categorisation of signal sources into audio and radio frequencies came about.

But there are many more applications for signal sources than just audio and radio frequency circuit testing. Servo systems, say, require a signal source varying between just a fraction of a hertz to just a few hertz – well below audio frequencies. Ultrasonic equipment needs a source of about 30 kHz – obviously above the audio frequency range, but just as obviously below that of radio frequencies. Microwave circuits, operating at frequencies up to around 10 GHz, would hardly warrant the name radio frequency, being of a somewhat greater frequency than the average Radio 4 transmission. The reader should also have noticed that all the applications listed so far here are basically analog in nature: but there is an increasing need in the testing of *digital* circuits to have signal sources capable of producing digital signals, too. **Logic pattern** or **word generators** are possible digital signal sources, merely producing

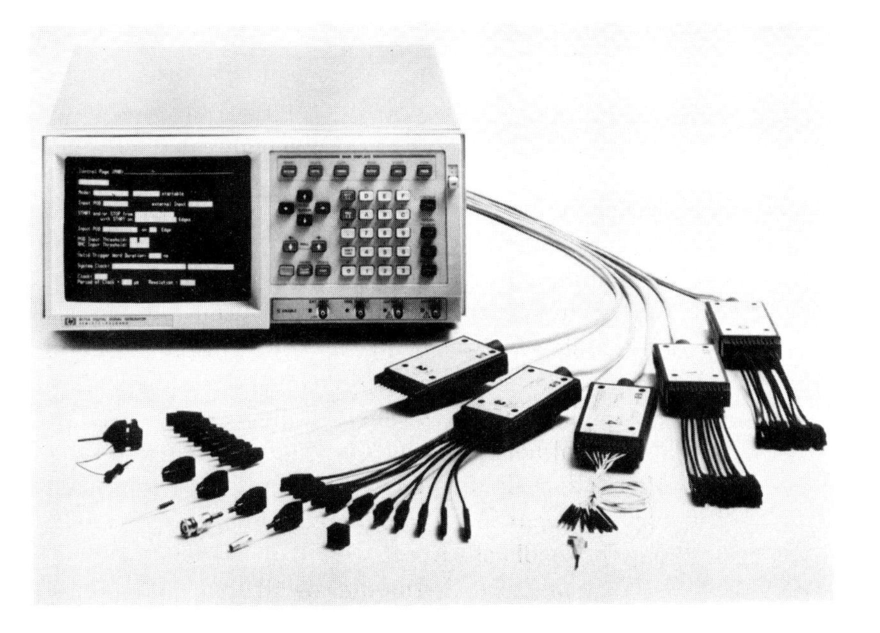

Photo 5.2 Hewlett Packard 8175A digital signal generator, shown with interface adaptor pods (Hewlett Packard)

serial or parallel data at a preselected speed. **Digital signal generators,** complete with cathode ray tube data display, are also available.

So, although we've noted the fact that signal sources have many more uses and functions than would seem apparent by their broad categorisation, we shall stick to the two traditional categories simply for convenience, pointing out inconsistencies as we proceed.

Low frequency oscillators

Although considerable differences exist in test equipment in this group, most low frequency oscillators are capable of producing signals in the frequency range of about 1 Hz to 1 MHz. Most instruments in this category cover this total frequency range in a number of switched ranges, allowing fine tuning within each smaller range with a potentiometer or similar variable device. Accuracy, stability of signal frequency, and stability of signal amplitude depend mainly on the techniques used for signal generation and attenuation.

The oscillator circuits traditionally used in low frequency oscillator test equipment are **harmonic oscillators;** that is, they produce sinusoidal output signals. In strict electronic terms, a harmonic oscillator is an amplifier which derives its input from its own output. Only part of the output signal is fed back to the input, the remainder is available as an output signal to following equipment. Figure 5.1 illustrates the basic principle of an oscillator. For the oscillator to function, two main criteria must be fulfilled. First, at some frequency f_0, the total phase shift caused by the amplifier and the feedback network must be zero. Second, at the frequency f_0, the amplifier gain must be sufficient to just compensate for the loss caused by the feedback circuit.

All practical harmonic oscillators fulfil these

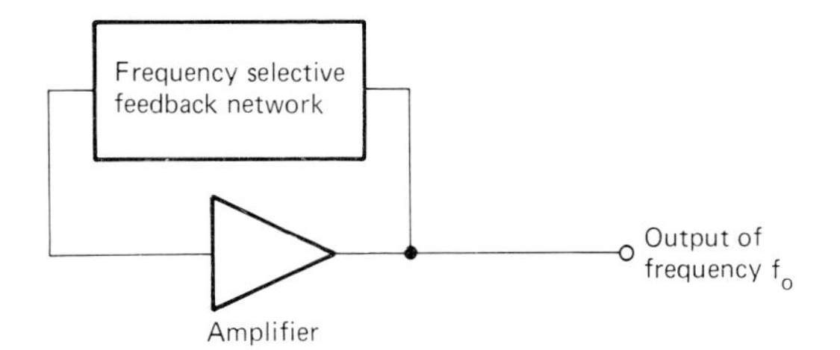

Figure 5.1 Block diagram showing the basic oscillator principle

criteria. At frequencies other than f_0, however, one or other of the criteria is not fulfilled and so oscillation cannot occur – in other words, the oscillator only oscillates at the particular frequency, f_0. A resistor-capacitor network whose phase shift and attenuation is dependent on frequency is used as the feedback circuit which, together with an amplifier, forms the oscillator. Three main types of resistor-capacitor networks are used in low frequency harmonic oscillators; the **Wien bridge network**, the **phase shift network**, and the **bridged-T network**, shown in Figure 5.2a, b and c. The Wien bridge type of oscillator is most common. By using a ganged potentiometer as the resistances of the feedback network, the oscillator frequency can be tuned over a limited range and, by switching a number of capacitors into and out of circuit, different ranges of frequency are obtained.

Typical distortion of harmonic oscillator test equipment is around 0.1 per cent but specialised equipment with distortion of only 0.001 per cent is available.

Function generators

In low frequency oscillator test equipment of a more modern design, **relaxation oscillators** are common. A relaxation oscillator is one in which one or more voltages or currents change suddenly during each cycle of the oscillation. There are a number of circuit possibilities, common examples being the astable multivibrator and the unijunction transistor oscillator;

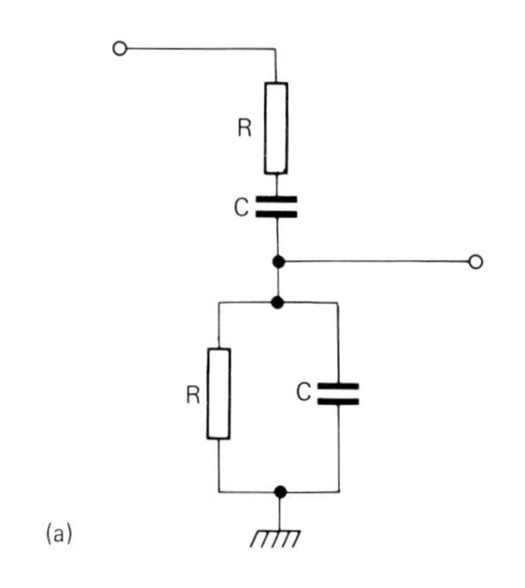

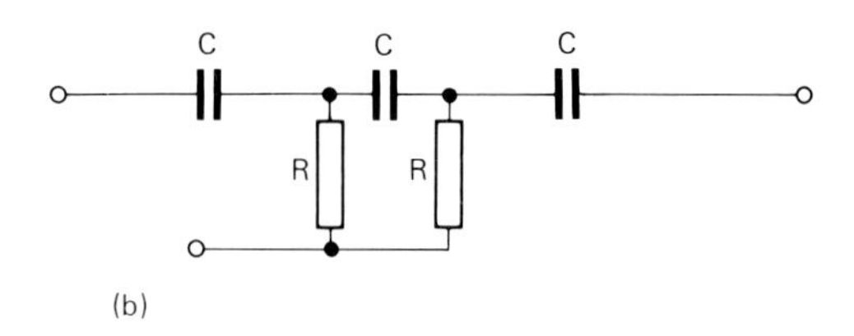

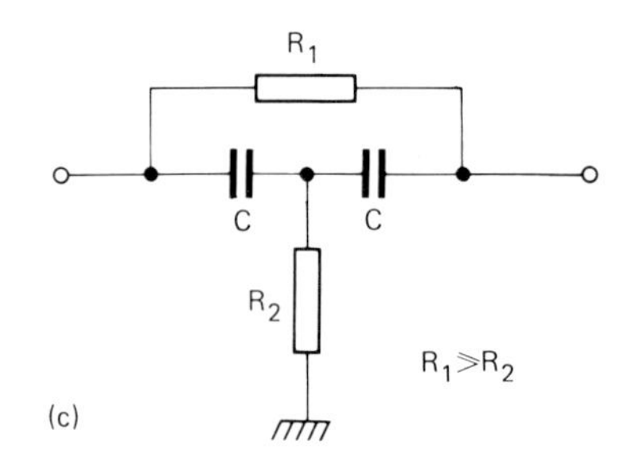

Figure 5.2 Three types of frequency selective feedback circuits used in oscillators (a) Wien bridge circuit (b) phase shift circuit (c) bridge-T circuit

but the design which is now most regularly used in low frequency oscillator test equipment is known as the **function generator**, a block diagram of which is shown in Figure 5.3. The heart of the function generator is a voltage-controlled squarewave oscillator which produces a squarewave signal at the required frequency. An integrator produces a triangular wave-

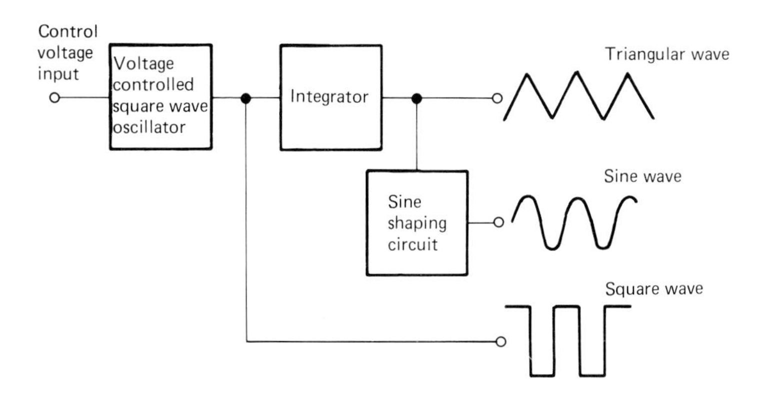

Figure 5.3 Basic principle of a function generator. In practice the voltage-controlled oscillator and the integrator interact more closely

form from this squarewave signal, and a sinewave-shaping circuit produces a sinusoidal-type waveform from the triangular waveform. This sinewave-shaping circuit can only approximate a true sinusoidal waveform, however, and distortion of around 1.5 per cent is common. Some high-quality function generators are also capable of producing sawtooth waveforms and pulse waveforms.

In a practical circuit the squarewave oscillator and the integrator are not truly separate blocks, as they closely interact. For example, the value of the integrator capacitor also determines the basic frequency range of the squarewave oscillator, so by switching into and out of circuit different capacitor values, different frequency ranges are covered by the function generator. Typical frequency range covered by function generators is from about 0.01 Hz to 2 MHz.

Fine tuning of the function generator's signal frequency is accomplished by varying the voltage at the voltage-controlled squarewave oscillator's control input. As frequency *is* voltage controlled it is a simple step to provide a second oscillator circuit within the function generator test equipment, which is used to sweep the oscillator frequency within a chosen range. A high-quality function generator's

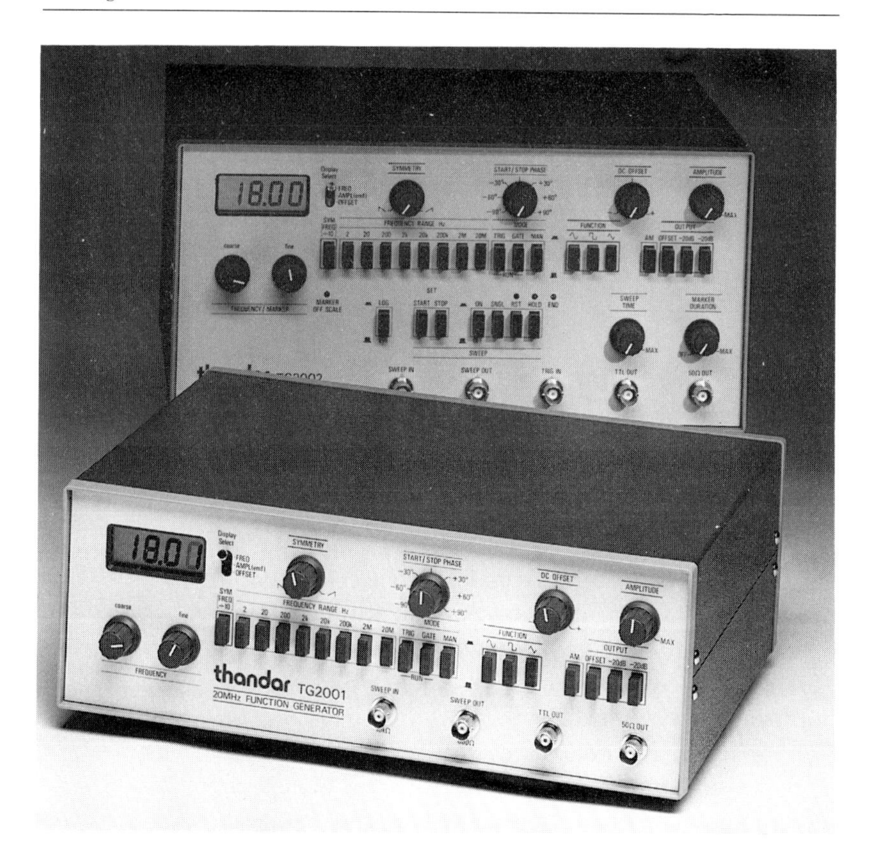

Photo 5.3 Thandar TG2001 and TG2002 function generators (Thurlby-Thandar)

second oscillator has a controllable frequency and amplitude allowing the user to control sweep range and speed, but medium quality generators only have a fixed speed second oscillator. Cheaper function generators merely have a sweep input terminal which allows the user to connect an external low frequency oscillator for this purpose.

Digital low frequency oscillators

In low frequency oscillators of this type, an analog waveform output signal is synthesised from stored digital information. The digital information corresponding to the required analog waveform is stored in a ROM device and read out step-by-step before being converted to an analog form by a digital-to-analog converter. The principle of operation is shown in Figure 5.4. Waveforms of any type, e.g. squarewave or triangular wave, may be stored in the ROM and read out as required. The digital information is initially stored in the ROM after sampling accurate examples of each required waveforms and making an analog-to-digital conversion of each sample.

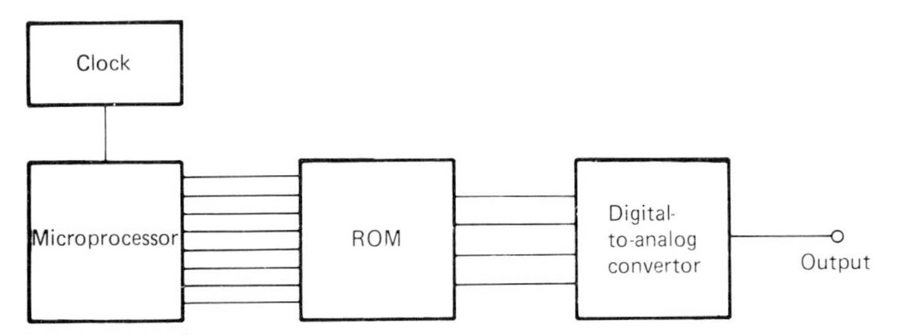

Figure 5.4 Digital synthesis of waveforms from samples stored in ROM

Choosing which waveform to be output from the digitally synthesised oscillator is simply a matter of reading the corresponding ROM locations of each waveform. The microprocessor accesses that part of ROM which contains the chosen waveform, and one location at a time is read in sequence with each clock cycle. Signal frequency is therefore determined by the frequency of the clock: if there are, say, 100 stored data words for each cycle of waveform, then signal frequency is equal to the clock frequency divided by 100.

This method of signal generation has a number of advantages. First, the clock can be frequency locked to a sub-multiple of a crystal reference oscillator, so that an oscillator output signal of accurately fixed frequency is obtained – accuracies of ± 0.0005 per cent are possible. Second, the clock frequency may be voltage controlled, allowing extensive sweep facilities with little extra cost. Third, amplitude stability is extremely good due to the digital generation techniques.

A disadvantage, on the other hand, is distortion. which is around 1 per cent and is due mainly to quantisation in the analog-to-digital and digital-to-analog processes. This can be reduced in two ways: by increasing the number of quantisation steps; and by using low-pass filters in the output stage to filter out distortion components. High-quality synthesised low frequency oscillators have typical distortion figures of less than 0.1 per cent.

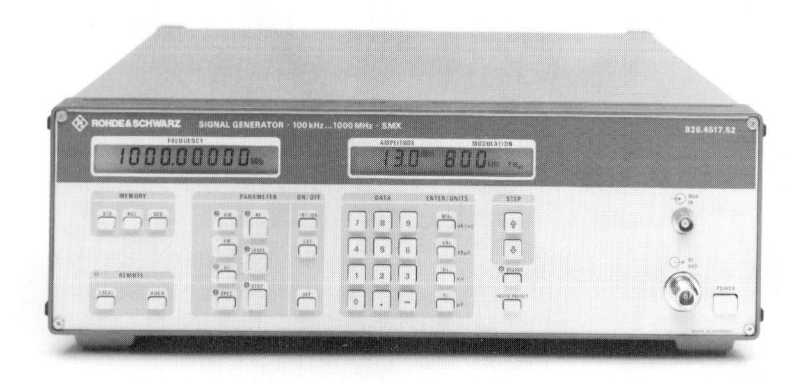

Photo 5.4 Rohde & Schwarz SMX signal generator (Rohde & Schwarz)

Signal generators

In the same way that reasonable quality but basic low frequency oscillators can be constructed with simple oscillator circuits, so it is that reasonable quality signal generators can be made. The two most common oscillator circuits used in such signal generators are the **Hartley oscillator** and the **Colpitts oscillator,** shown in Figure 5.5a and b. Variable tuning capacitors allow the oscillator to be tuned only over about a 3:1 range and so much switching in and out of circuit of capacitors and inductors is required. Although harmonic distortion of the generated sinusoidal waveform is quite low with these

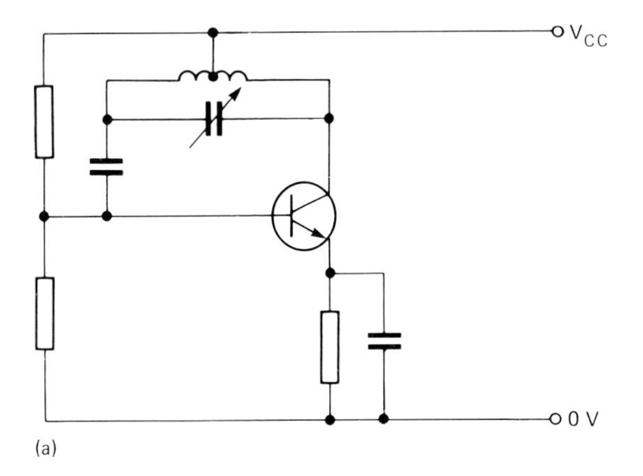

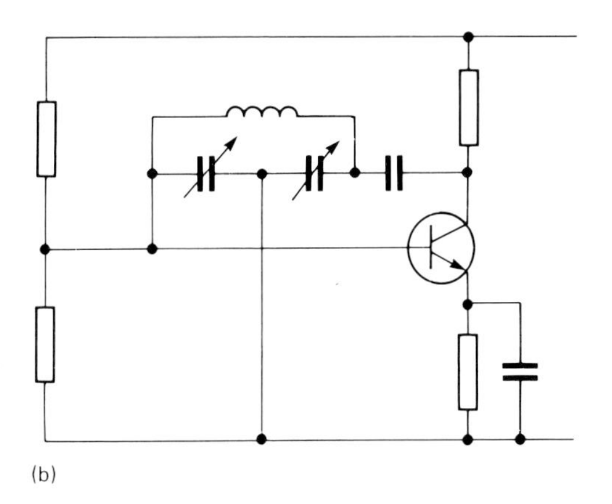

Figure 5.5 Two oscillator circuits forming the basis of common signal generators (a) Hartley oscillator (b) Colpitts oscillator

oscillators, their frequency stability is not particularly good.

The use of varicap diodes (an abbreviation of *variable capacitance*), semiconductor devices that exhibit junction capacitances which vary dependent on their junction voltages, to replace tuning capacitors in the basic oscillator circuits means that the frequency of oscillation can be controlled electronically. This is not particularly an advantage for signal generators using the basic principles of oscillation, but *is* an advantage in more complex signal generators, as we shall see.

For some specialised applications basic oscillators

Photo 5.5 Analogic 2020 formula-input waveform synthesiser (Dryden Brown)

cannot produce the required frequency range, and the user may wish to go to the expense of a **cavity-tuned oscillator,** using a resonant cavity. Strictly speaking, resonant cavities are only capable of producing microwave frequency signals, but the device used to generate radio frequency signals for cavity-tuned signal generators – a capacitor-inductor device with variable capacitance – is sufficiently cavity-like to warrant the name. The device is capable of generating signals of frequencies from about 10 kHz to over 1000 MHz, in a single range.

Heterodyne oscillators

The narrow tuning ranges possible with the basic oscillators above are an obvious disadvantage. A **heterodyne oscillator** can beat this problem, without the expense of a cavity-tuned oscillator. A block diagram is shown in Figure 5.6, where we can see that the output from a stable, fixed frequency oscillator is mixed with the output from a variable

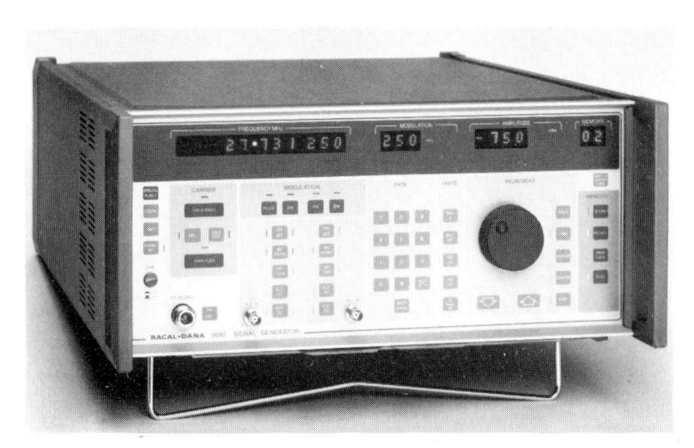

Photo 5.6 Racal-Dana 9087 synthesised signal generator (Racal Group Services)

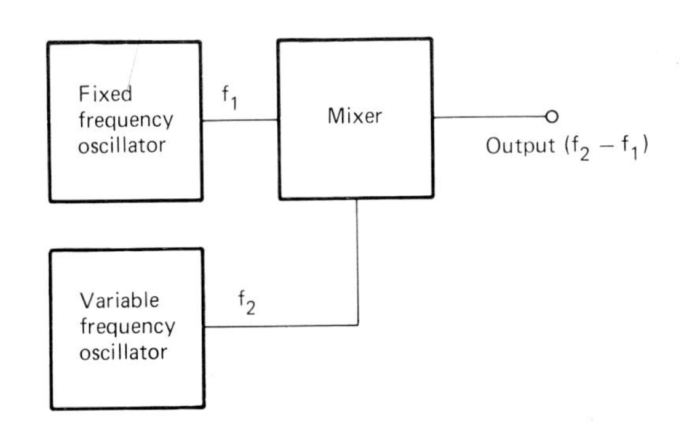

Figure 5.6 Block diagram of a heterodyne oscillator

frequency oscillator. The output frequency of the signal generator is the difference:

$$f_2 - f_1$$

where f_1 is the fixed oscillator frequency, and f_2 is the variable oscillator frequency. Thus, if f_1 is, say, 200 MHz, and f_2 is, say, variable between 200.1 MHz and 300 MHz (perfectly possible with a Colpitts or Hartley oscillator), then the frequency range of the heterodyne oscillator signal generator is from 0.1 MHz to 100 MHz – a vast improvement over the 3:1 tuning range above.

Heterodyne oscillators, however, typically have poor frequency stabilities as they rely on the frequency stability of the variable oscillator, and they produce a significant amount of noise and spurious signals.

Phase-locked loop oscillators

A more common method of radio frequency oscillation in modern signal generators is the **phase-locked loop**, sometimes called the **synthesiser** method, illustrated in Figure 5.7. In this type of signal generator a fixed frequency and highly stable reference oscillator (usually a crystal reference) is used. A phase detecting circuit compares the phase of this reference oscillation with the phase of the signal from a voltage-controlled oscillator (of the varicap type already discussed) after it has first passed through a variable divider circuit. The voltage-controlled oscillator's control voltage is supplied by the phase detector.

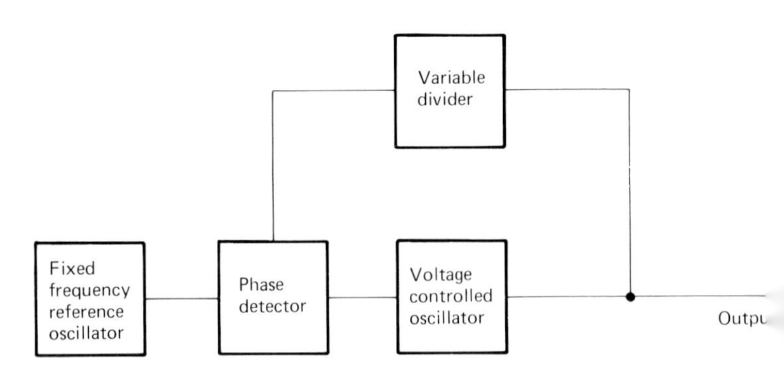

Figure 5.7 Principle of a phased-locked loop, or synthesiser oscillator

If the reference oscillator signal differs in phase from the divided voltage-controlled oscillator output, the controlling voltage is adjusted to bring the two back into phase, thus locking the voltage-controlled oscillator's output signal to a stable frequency. The feedback loop formed by the phase-locking principle stabilises the voltage-controlled oscillator to a frequency equal to the reference oscillator frequency multiplied by the division ratio of the divider circuit, so by changing the division ratio different frequencies of output signal can be generated. These frequencies are, however, in fixed steps, dependent on the division ratios.

Some modern designs of synthesised signal generators use microprocessor-controlled divider circuits, which allow tuning in very small steps (typically 10 Hz resolution, or less) over a very wide range of frequencies (say, 100 kHz to 1000 MHz). Some latest designs of synthesised signal generators use the microprocessor to control the synthesised output, in a functional form $y = f(t)$. The user of such **formula-input waveform synthesisers** merely inputs the function in this form, via front panel keys, to produce the required output signal. Top-of-the-range synthesised signal generators have a frequency range from about 10 kHz to over 1 GHz, with best resolutions of around 1 Hz.

General output requirements

Modulation of the generated signal can occur in a number of ways. In low frequency oscillators an external or internal sweep oscillator is used to modulate the signal frequency. In this application, the so-called sweep is just another name for frequency modulation, and in fact, radio frequency signal generators stick to the correct term. As we'll see later, there is another meaning of the term 'sweep'.

Frequency modulation is accomplished in most oscillators by modulating the control voltage to the varicap diode. Amplitude modulation of the oscillator signal is generally accomplished after the oscillator stage, with a logically, though not necessarily physically, distinct modulator circuit. In low frequency oscillators, amplitude modulation can be accomplished using a voltage-controlled amplifier. Pulse modulation and sometimes phase modulation are possible requirements of a signal source, too.

A signal source's output signal is not usually taken directly from the oscillator or amplitude modulator. Generally an amplifier is used to provide sufficient output level and to buffer the oscillator against load impedance changes. Switched attenuators of some kind are usually included to provide low level signals where required.

Particularly in signal generators, but also to an extent with low frequency oscillators, output level amplitude must be accurately maintained. Automatic level control systems, constantly monitoring the output level and adjusting accordingly in a feedback loop, are usual. These circuits also allow output level to be varied over a limited range. In combination with the switched attenuators mentioned in the previous paragraph a large range of signal levels is therefore available. Generally, output level stages are calibrated so the user can accurately choose signal level. Some way of metering the output signal level and frequency is useful and analog meters can be used for this purpose, but in the case of synthesised oscillators, particularly of the microprocessor-controlled kind, it is a relatively simple task to provide a digital metering display.

Two further requirements of output stages of signal generators are good screening of attenuator stages to prevent interference from higher level signals elsewhere in the equipment, and reverse power protection, that is, protection against damage by signals generated in other equipment connected to the signal generator.

Sweep generators

A rather obvious extension for the signal generator is as a **sweep generator,** a sort of frequency modulated

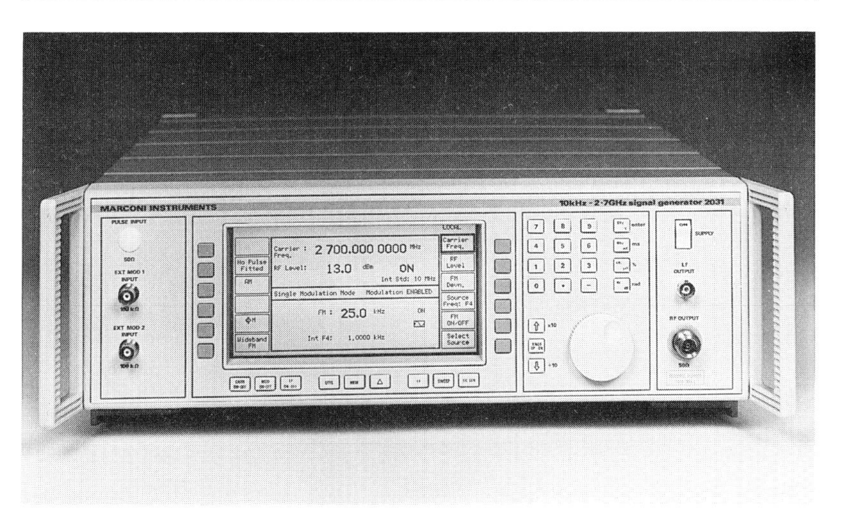

Photo 5.7 Marconi 2031 signal generator (Marconi Instruments)

signal generator in which the amount and range of frequency modulation is controlled by an accurately produced time-base. Sweep generators are traditionally used to measure the frequency response of circuits, by supplying the signal which is swept across the circuit's range. The circuit's output signal is then displayed on the cathode ray tube of an oscilloscope as a graph of amplitude against frequency or, in the case of some sweep generators, on the sweep generator's own internal display.

Figure 5.8 Block diagram of a sweep generator

Figure 5.8 shows the main parts of a sweep generator as a block diagram.

The timebase is much like that of an oscilloscope, but does not need to be as complex; relatively few speed variations are needed and a fast flyback is not essential. Some sort of control is required between the timebase circuit and the voltage-controlled oscillator, however, to adjust the timebase voltage so that a chosen range of frequencies may be generated. This frequency controller also defines a number of possible operating modes, the main two being: the start-stop mode (also called the f_1-to-f_2 mode), where the sweep start and stop frequencies are set with two separate front panel control knobs; and the delta frequency mode (Δf mode), in which the sweep centre frequency is set with one front panel control knob and the deviation around that frequency set with another.

Frequency marker circuits allow markers to be superimposed on the display, generally at regular intervals, thus calibrating the display with a set of reference frequency points, commonly called *birdies*. These circuits are fairly simple harmonic generators, producing harmonics of a stable crystal-controlled oscillator's output.

The voltage-controlled oscillator can be of any type but, as a sweep generator is usually required to sweep over a number of frequency decades, there is a problem with simple oscillators, which have limited tuning ranges. One way of overcoming the problem is to divide the total range into a number of bands and electronically switch the bands as the sweep proceeds. The principle of such *stacking* sweep generators is illustrated in Figure 5.9. Heterodyne oscillators or synthesised oscillators are capable of producing the required frequency ranges, however, without the necessity of band switching.

An additional feature found on many sweep

Photo 5.8 Marconi 2022C signal generator (Marconi Instruments)

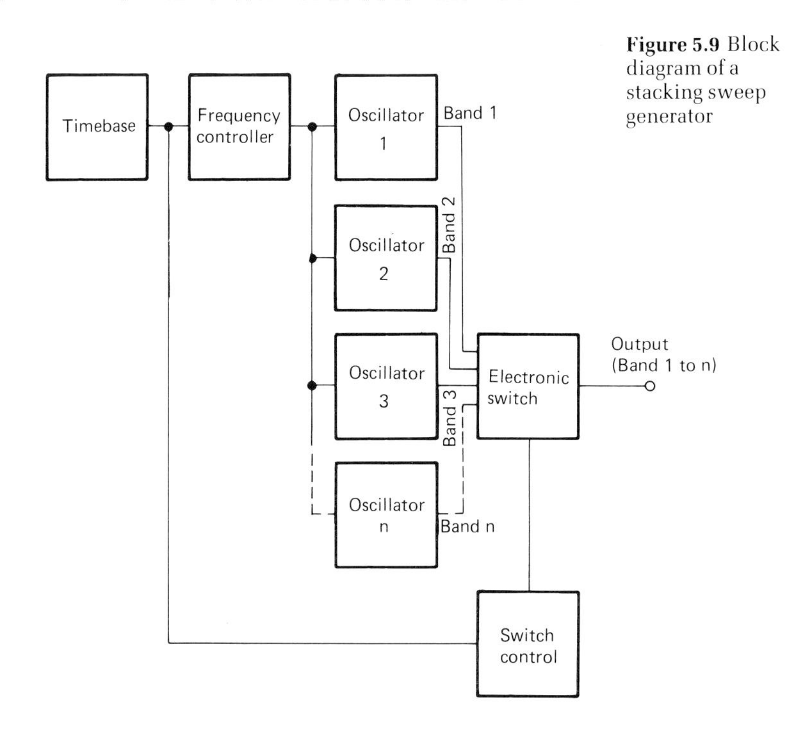

Figure 5.9 Block diagram of a stacking sweep generator

generators is the facility to change from a linear to a logarithmic display of signal level information. If large changes in level are being observed, the logarithmic display gives best presentation, but for small changes in level the linear display is best.

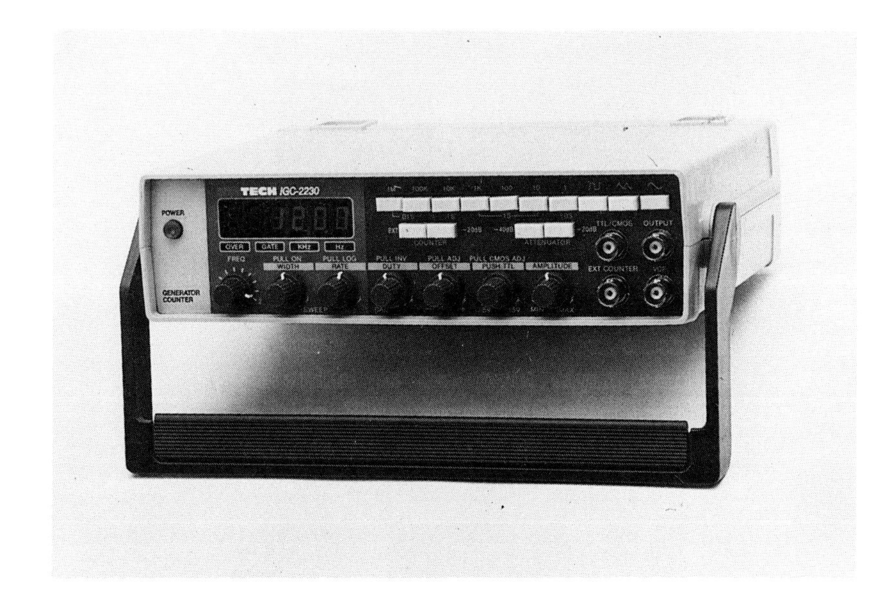

Photo 5.9 A function generator/counter (RS Components Ltd)

6
Frequency, time and event counters

Frequency, time and event measurements are accomplished with the use of test instruments with the generic name **counters,** sometimes called **universal counter timers** (UCTs). Counters are generally digital in operation and so have a digital read-out display of LED, LCD, vacuum fluorescent or similar form.

Operation of all counters depends on the principle of gating a signal over a time, while counting the number of pulses of the signal during that time. Figure 6.1 illustrates the principle where a signal of unknown frequency is gated for a specified period of time, during which a number of pulses of the signal are counted and displayed. If, say, the signal is of a frequency of 500 Hz, and the gating period is 1 second, then the counted and displayed results is 500 Hz.

It doesn't take long to figure out that the gating

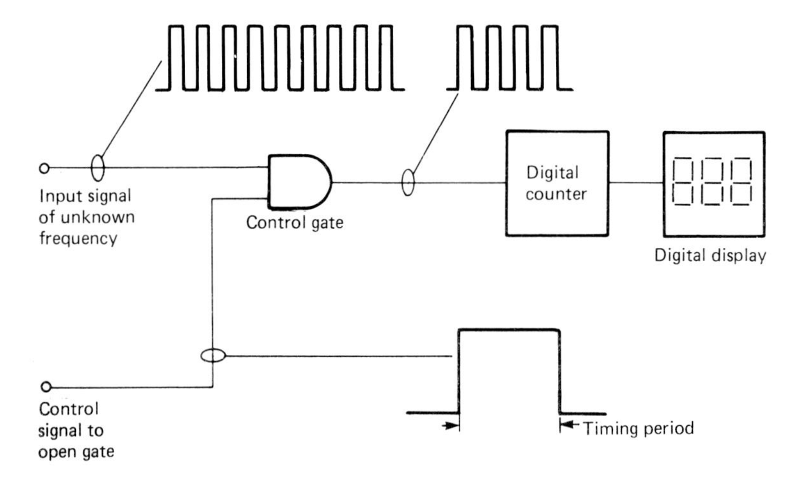

Figure 6.1 Basic direct gated counter principle. The input signal is counted over a period of time

period is quite critical. If, say, the gating period in this example is 1.1 s (a difference of 10 per cent), then the displayed result is 550; an error of 10 per cent. In **direct gated counters** of this form the gating period is normally achieved by dividing down the output signal of a stable high frequency reference oscillator, sometimes called the timebase or clock, as shown in the block diagram of Figure 6.2, to give accurately controlled time periods. Also shown in Figure 6.2 is a signal amplifier and conditioner block, which accepts input signals from a wide range of sources and conditions them to a pulse-type form which the digital counter block can directly count.

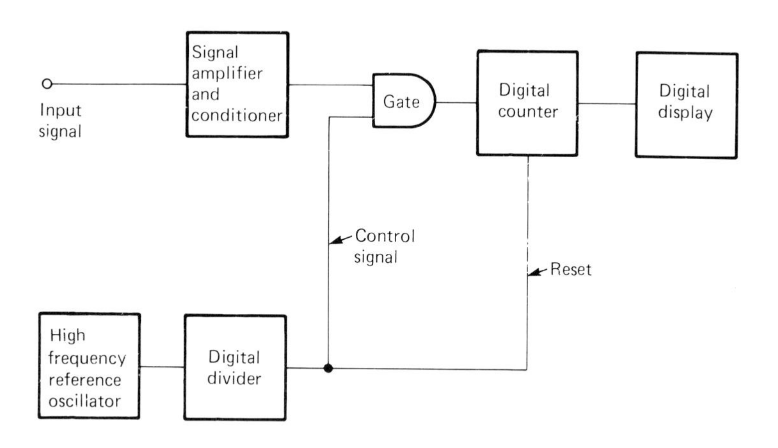

Figure 6.2 A high frequency reference oscillator and divider are used to define an exact timing period

High frequency measurement

Direct gated counters are very accurate within the limitations of the method. They are particularly useful for measurement of reasonably high frequency signals or events – direct gated counters measuring signals over 500 MHz are common – and, indeed, the upper limit of measurement depends on the speed of the logic circuits in the digital counter, not on the accuracy of the gating period. For even higher

Photo 6.1 Keithley 775 programmable counter/timer (Keithley Instruments)

frequencies, however, the direct gating method may be adapted in a number of ways. First, **prescalers** can be included, as shown in Figure 6.3, either within the counter or normally as an extra add-on device, to divide down the input signal frequency to a frequency which the direct gated counter can measure. Prescalers are simple dividing circuits, and as such do not have the high frequency limitations of digital counters. In use, the gating period of the counter must be made longer, by the same factor as the prescaler reduces the input signal.

Figure 6.3 Use of a prescaler to measure higher frequency signals than the digital counter is capable of

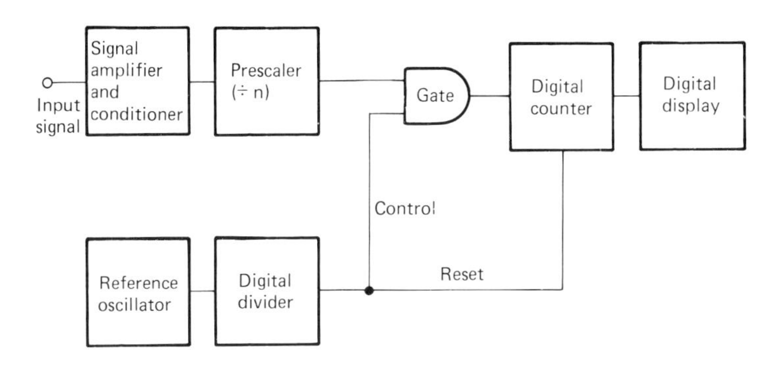

Another method of adapting direct gated counters to measure higher frequencies is by heterodyning the input signal down to a lower frequency, illustrated in Figure 6.4. The resultant signal is counted during a gating period which is dependent on the heterodyning signal, then displayed as usual. **Heterodyne counters** are much more complex than other types, but have the advantage that they can measure input signal frequencies of over 20 GHz. Even higher frequencies may be measured using an adaptation of the heterodyne counter, the **synthesised heterodyne counter,** which uses a microprocessor to form a synthesised oscillator timebase and control the measurement procedures. The resultant test equipment will measure up to about 40 GHz and is highly specialised.

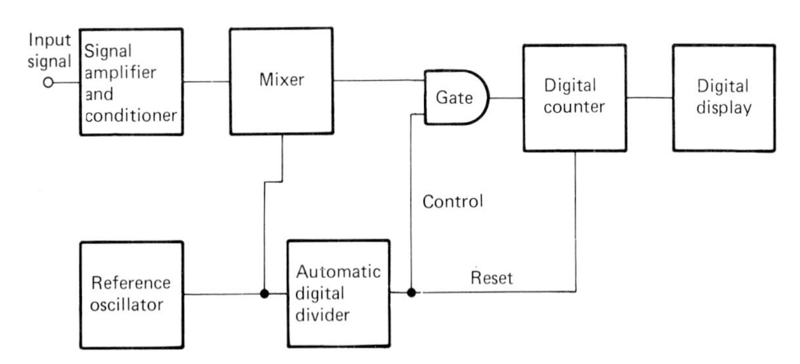

Figure 6.4 Block diagram of a heterodyne counter

Low frequency measurement

Although direct gated counters can measure reasonably high frequency signals, they are not so useful for low frequency measurements, where accuracy depends largely on the length of time the user is prepared to wait for counting and display to take place. A frequency of 10 Hz, for example, needs to be counted over a period of 100 s for an accuracy of ± 1 per cent.

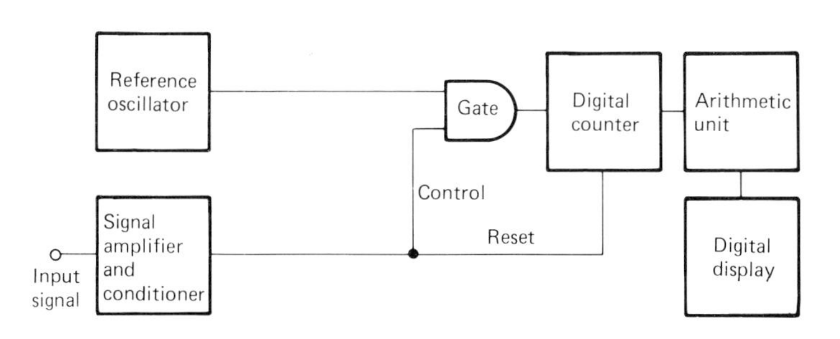

Figure 6.5 Basic block diagram of a reciprocating gated counter used to measure low frequency signals

Low frequencies can in fact be measured using the same principle but with one major adaptation: by gating the timebase with the input signal, as shown in the block diagram of Figure 6.5. An input frequency of, say, 1 Hz gates the timebase frequency for a 0.5 s interval every second, so with a timebase frequency of, say, 2 MHz a count of 1 000 000 pulses is made. The reciprocal of this count is calculated by an arithmetic unit and the result, 0.000 001, is displayed with a MHz scale. Now, 0.000 001 MHz is, of course, 1 Hz. Counters of this type are called **reciprocating gated counters.**

Another method of measuring low frequency events is with the inclusion of a **phase-locked loop** circuit to multiply the frequency of the input signal, before gating, counting and displaying the result. A block diagram of a **multiplying counter** using a phase-locked loop multiplier is shown in Figure 6.6. The voltage-controlled oscillator's output frequency

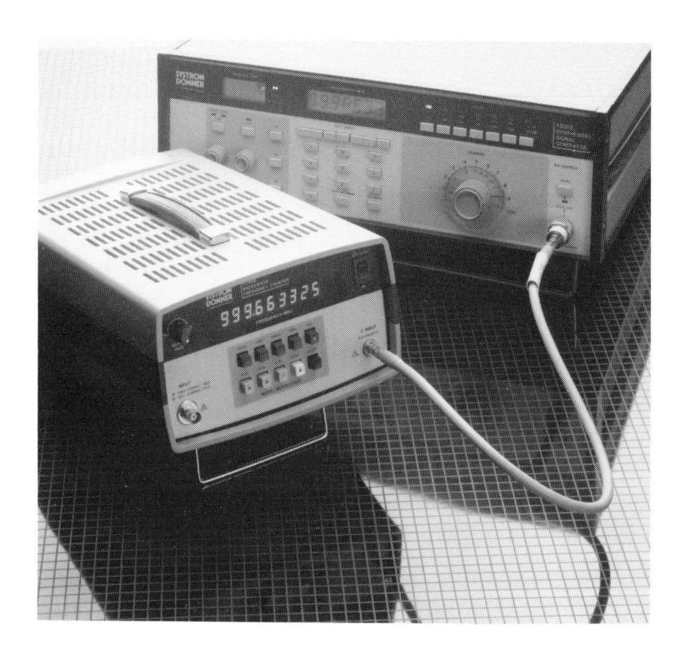

Photo 6.2 Systron Donner 6246B microwave frequency counter, shown in use with the 1300 synthesised signal generator (Thorn EMI Instruments)

is equal to the input signal frequency multiplied by the divider circuit's division ratio.

Equipment and operation

Most counters aren't formed by just *one* of the previous counting methods and, generally, are a combination of two or more. General-purpose counters are typically able to measure frequencies from DC through to about 500 MHz, so they are combinations of the direct gated and either the reciprocating gated or multiplying methods. Counters able to measure higher frequencies will usually maintain this combination, merely adding a third, or even a fourth method. Front panel controls determine which measurement method is in operation, although higher frequency measurements are accomplished using a totally different input terminal from the lower frequency measurement inputs.

Although so far in this discussion we have been

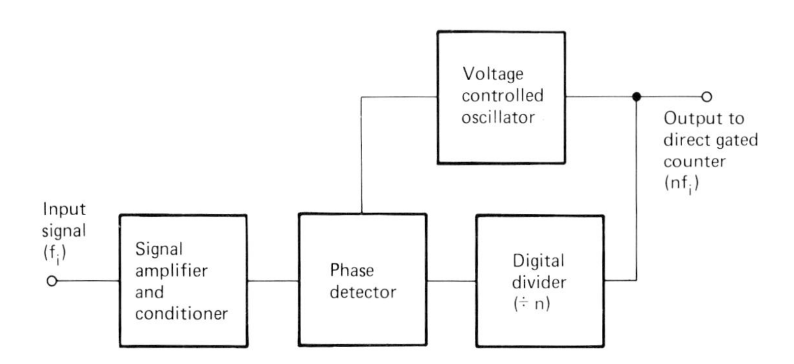

Figure 6.6 Principle of a multiplying counter using a phase-locked loop to increase the frequency of a low frequency input signal before direct gated counting

looking at the counter's ability to count and display a measurement of input signal frequency, counters are capable of much more than this. Typically, a counter has two general-purpose input channels, each with its own input terminals and normally marked Channel A and Channel B. These channels are identical in performance and operation, and frequency measurements of input signals can be made using either channel. Various switches or controls allow user-control of signal conditioning; in particular, signal attenuation and trigger level – these are much like the same controls on oscilloscopes. The circuits corresponding to channels A

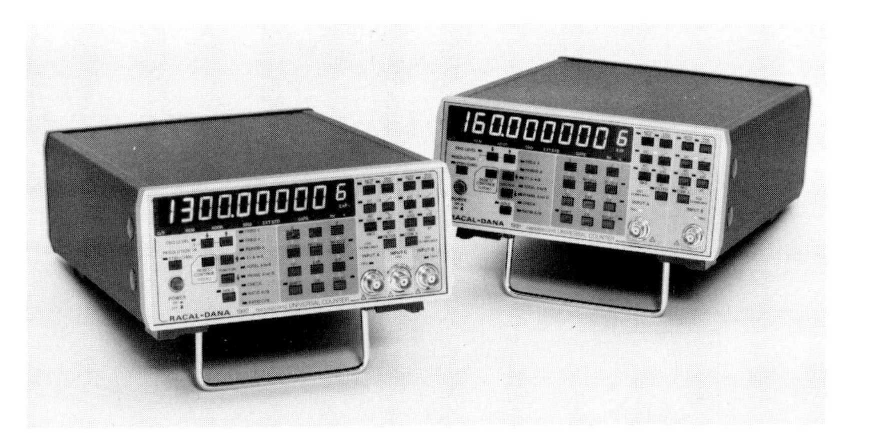

Photo 6.3 Racal-Dana 1991 and 1992 universal counter/timers (Racal Group Services)

and B are generally direct gated counters, combined with reciprocating gated or multiplying counters, so that either channel may be used to measure frequencies from DC to a maximum of, say, 500 MHz. If the counter is capable of measuring higher frequencies a third circuit, Channel C, is included which operates using one of the higher frequency adaptations of the direct gated counter. Channel C is used to measure input signal frequencies of, say, 100 MHz through to the counter's maximum. Usually, though not necessarily, Channel C is physically separate from the other two channels within the counter – this allows the equipment manufacturer to provide adequate screening against the possibility of high frequency interference.

In frequency measurements, the counter gate is controlled automatically by the internal timebase. By allowing the gate to be controlled externally and counting the timebase signal pulses, however, a measurement and display of the *time interval* between opening and closing the gate is made. This is illustrated in Figure 6.7. If, say, the start and stop control signals are 1 s apart, and the timebase is of

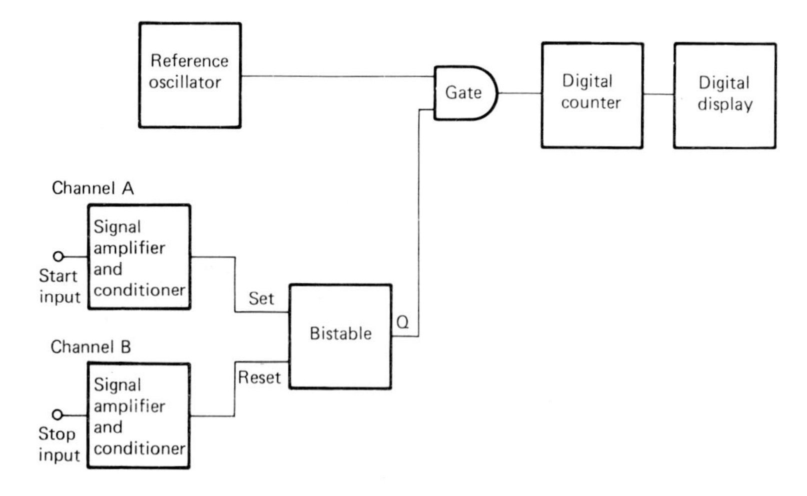

Figure 6.7 Using the two input channels of a counter to measure the time interval between two input pulses

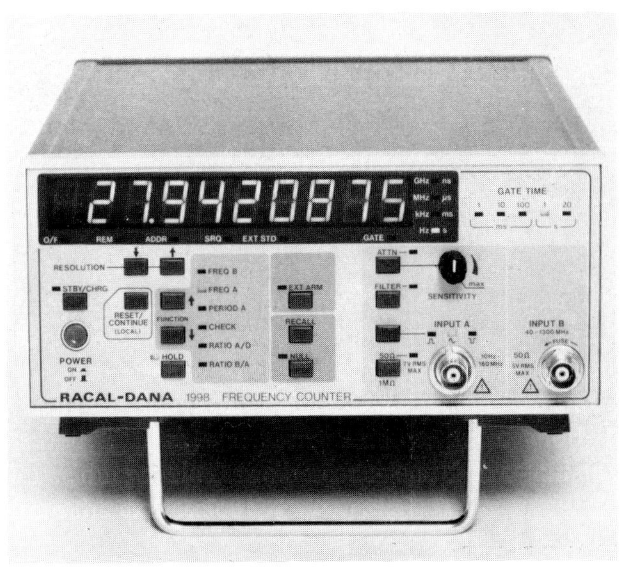

Photo 6.4 Racal-Dana 1998 frequency counter (Racal Group Services)

1 MHz, then a count of 1 000 000 is made. Simply by positioning the display's decimal point correctly, a display of 1.000000 shows the time interval between start and stop signals to be 1 s. In start-stop time measurements of this kind, a received pulse to Channel A is used to open the gate, and a received pulse to Channel B closes the gate – hence the importance of two identical input channels.

Event counting is accomplished by opening the counter gate, manually with a front panel control or remotely with a pulse input, and letting the counter simply add the received signal pulses to Channel A. A received pulse to Channel B closes the gate, and the counter displays the number of pulses received at Channel A during the period between opening and closing the gate. The timebase itself is neither counted nor used to gate the counter in this mode.

Errors

In frequency and time measurements the primary cause of measurement error is the accuracy with

Photo 6.5 Marconi 2440 microwave counter (Marconi Instruments)

which the timebase maintains its frequency of oscillation. Whatever error this reference oscillation has is directly reflected in the measurement itself. Most counters generate the timebase with a quartz crystal oscillator of one type or another, and even the simplest crystal oscillator produces acceptable results in many applications. The main reasons for timebase inaccuracies are that crystal characteristics are temperature dependent and age dependent; that is, oscillator frequency may vary with temperature and age. Small frequency variations also occur with changes in power supply voltage.

One way to reduce temperature variations is to use a resistor network to produce a voltage change dependent on temperature change. This voltage change is used to adjust the frequency of the crystal oscillator in a feedback-type loop. Such **temperature compensated crystal oscillators** (TCXO) are complex and require microprocessor-controlled circuit techniques.

The most accurate crystal oscillators mount the crystal in a temperature-controlled oven. Resulting **oven controlled crystal oscillators** (OCXO) have extremely low temperature coefficients. Oven mounting techniques can also improve the crystal ageing rate, too, because the crystal temperature is more constant.

Where internal timebases prove to be unacceptable, most counters have the provision for an external timebase terminal, so that a standard reference frequency may be used instead. Many such standards are broadcast and are thus freely available for users with the necessary receiving equipment.

Errors also arise in frequency and time measurements due to an inherent reading error, caused by non-synchronous signals at the counter gate. Figure 6.8a illustrates an input signal together with the gate signal generated by the timebase. A count of five is made during the time the gate is open. Figure 6.8b shows the same signals with the input transposed slightly in time. The count is now four during the gate period. This ± 1 error is always present in direct gated frequency measurements, and always affects the least significant digit of the count; that is, a count of 1 000 000 will have an error of ± 1 part in 10^6 while a count of 10 has an error of ± 1 part in 10. The

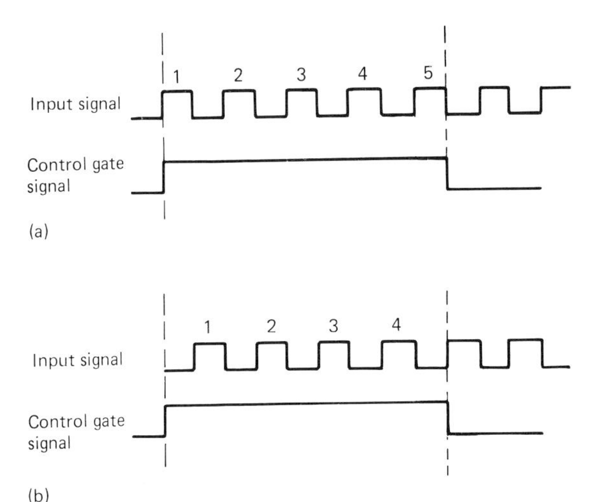

Figure 6.8 Counter reading error inherent in all direct gated frequency measurements

error as a percentage is thus dependent on the value of the count.

Time measurements suffer from a further error: trigger error, caused by noise on the input signal triggering the counter before or after it would have been triggered without the noise. It may be calculated from:

$$t_e = \frac{2 \times \text{Peak noise voltage (V)}}{\text{Signal slope at trigger point (V}\ \mu\text{s}^{-1}\text{)}}$$

where t_e is the trigger error, in μs.

Trigger error is related to the signal-to-noise ratio of the input signal and, in fact, for a signal-to-noise ratio of 40 dB trigger error is taken to be ± 0.3 per cent of the measurement period.

7
Spectrum analysers

According to Fourier analysis, all electrical waveforms can be thought of as a combination of sinusoidal signals of various frequencies and amplitudes. If these sinusoidal frequencies are related, so that the overall waveform is periodic the waveform can be displayed on an ordinary oscilloscope in the **time domain;** that is, the usual amplitude against time graph form. If the sinusoidal signals are unrelated, however, the oscilloscope cannot produce a stable display of the resultant unperiodic waveform and test equipment which produces a display in the **frequency domain,** that is, an amplitude against frequency graph form, is usually resorted to. The **spectrum analyser** creates a frequency domain display which resolves the electrical waveform into its constituent frequency components and displays them individually. Apart from being able to display composite waveforms which the oscilloscope cannot, the spectrum analyser has many other uses.

Figure 7.1 illustrates the difference between time domain displays and frequency domain displays using harmonically related sinusoidal components. A collection of sinusoidal components at frequencies f, $3f$, and $5f$ are first illustrated in a three-dimensional form, where amplitude, time and frequency of the components are measured along the three axes. These components have amplitudes in the ratio 1.273:0.4244:0.2546. Sinusoidal components of these amplitude and frequency relations may be shown, by Fourier analysis, for form a single waveform approximating a squarewave.

When viewed in direction X of Figure 7.1, the

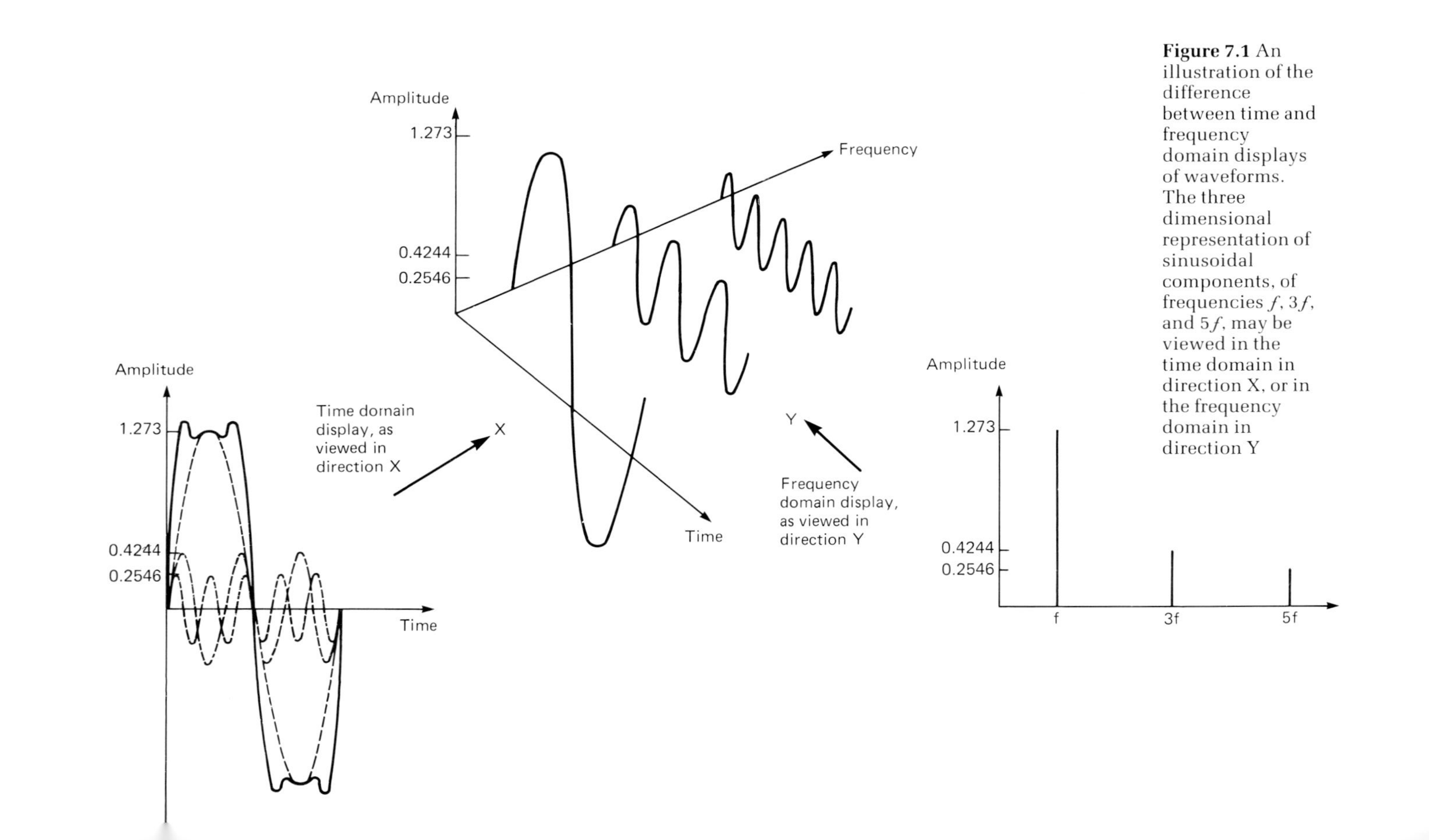

Figure 7.1 An illustration of the difference between time and frequency domain displays of waveforms. The three dimensional representation of sinusoidal components, of frequencies f, $3f$, and $5f$, may be viewed in the time domain in direction X, or in the frequency domain in direction Y

three components combine to create a composite periodic waveform of approximately squarewave form, with axes of amplitude (Y-axis) and time (X-axis). This is a time domain display of the resultant waveform, as would be displayed by an oscilloscope.

Viewed in direction Y, on the other hand, the three individual components are seen, as functions of amplitude (Y-axis) and frequency (X-axis). This is a frequency domain display of the resultant waveform, as would be displayed by a spectrum analyser.

Displays of frequency domain analysed waveforms are often more useful than those of time domain analysed waveforms, because individual components sometimes provide a greater understanding of what is going on. An example is distortion: when a distorted waveform is viewed in the time domain, no exact knowledge (apart from the obvious fact that it is distorted) of the type or cause of distortion can be calculated. In the frequency domain, however, an exact representation of all the waveform's components – including those, in this case, causing the distortion – is obtained.

Spectrum analyser types

There are four basic methods used by spectrum analysers to produce frequency domain displays. A

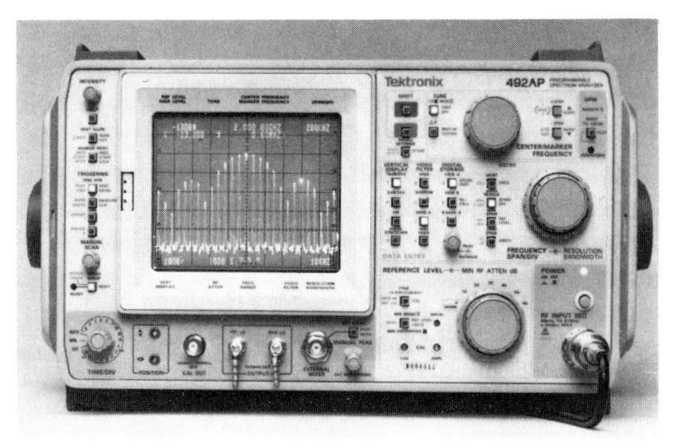

Photo 7.1 Tektronix 492AP programmable spectrum analyser (Tektronix UK)

real-time spectrum analyser comprises a number of bandpass filters whose responses are aligned such that the upper corner frequency of each filter intersects with the lower corner frequency of the next filter, as illustrated in Figure 7.2a.

The filter outputs are scanned in turn by an electronically controlled analog switching mechanism and used to define the controlling voltages applied to the vertical deflection plates of a cathode ray tube, as shown in Figure 7.2b. A scan generator which controls the analog switching mechanism also provides the horizontal deflection plates' controlling voltages. As each filter is, in turn, scanned and produces a

Figure 7.2 Principles of a real-time spectrum analyser (a) showing how each bandpass filter's response aligns with those of its neighbours (b) in block diagram form

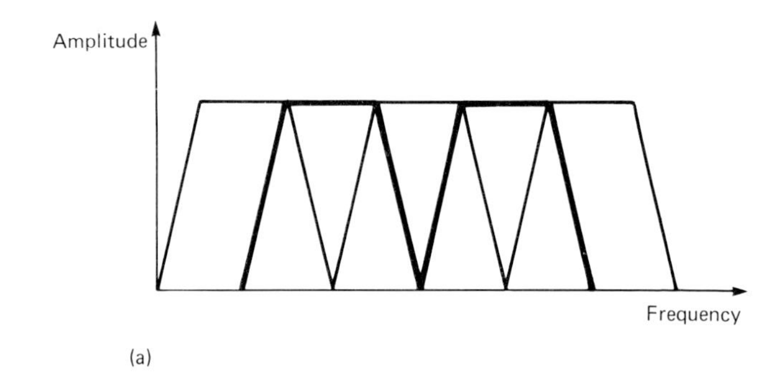

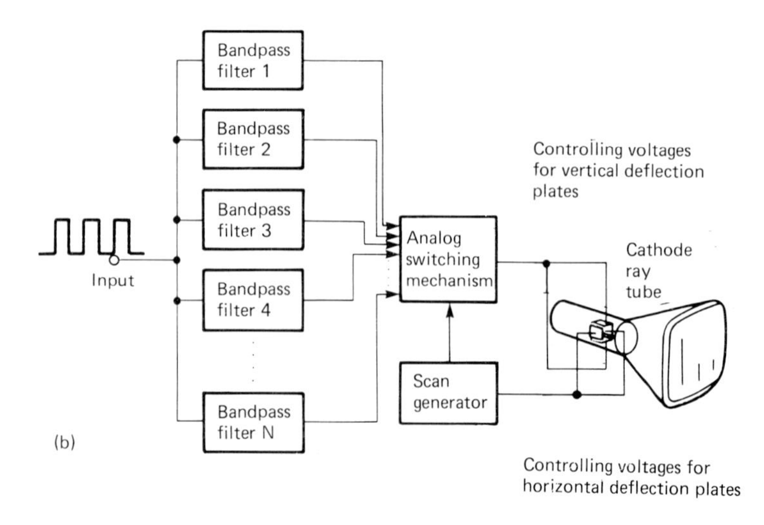

vertical beam deflection, the scan generator produces a corresponding left-to-right horizontal beam deflection, so that the resultant displayed waveform is a collection of signal amplitudes, one for each bandpass filter, across the screen.

Real-time spectrum analysers get their name from the fact that rapidly occurring events can be detected and displayed more or less as they occur, as the input signal is applied to all of the filters, all of the time. In some situations this is an advantage: transients, for example, are easily displayed. However, a corresponding disadvantage is the number of filters required. To produce a meaningful frequency domain display the filters must each have only a small bandwidth, otherwise signals with small variations of frequency spacing will appear merged on the display. An analogy may be taken with lines drawn with different thicknesses of pen tips. Where two lines are drawn with a fine-tipped pen, as in Figure 7.3a, the lines are close together but clearly seen. In Figure 7.3b, on the other hand, both lines have been drawn with a broad-tipped pen and have merged so that it is impossible to detect that more than one line exists. In a spectrum analyser the thickness of the 'line' produced is known as the **resolution**

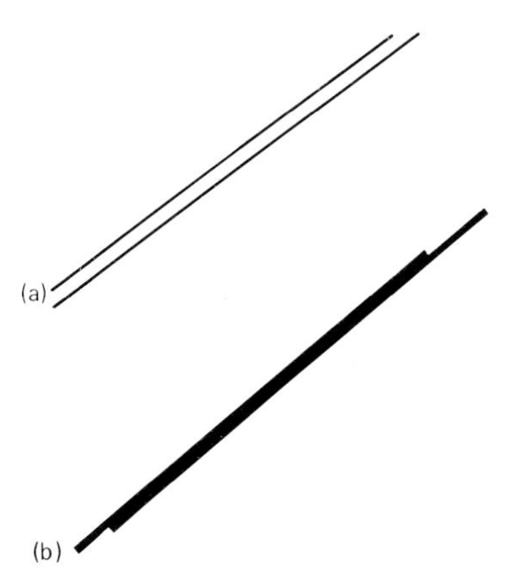

Figure 7.3 An analogy of the effects of resolution bandwidth (a) two lines drawn with a fine-tipped pen are clearly distinguishable (b) two lines drawn with a broad-tipped pen appear to merge and may not be distinguishable

bandwidth, and the smaller this is, the greater the analyser's capability to detect and display individual signals; that is, its resolution. In a real-time spectrum analyser, the resolution bandwidth is defined precisely by the bandwidth of each filter and the smaller the filter bandwidth the greater the analyser resolution. So, to create a spectrum analyser with an adequate resolution a very large number of accurate filters is required. Generally, for this reason, real-time spectrum analysers are quite expensive and are made for use only over limited frequency ranges, typically the audio range of around 0 Hz to 20 kHz. The subject of resolution bandwidth will be returned to, later.

An adaptation of this principle gives the second basic method of spectrum analysis and is shown in Figure 7.4, the **swept tuned spectrum analyser,** in which a single tunable bandpass filter is used to sweep over the whole frequency range covered by the analyser. This technique, of course, does not warrant the real-time classification of the previous one, as any rapid change in a frequency component of the input signal, will only be displayed if the filter is tuned specifically to that frequency – and most of the time the filter is tuned to other frequencies.

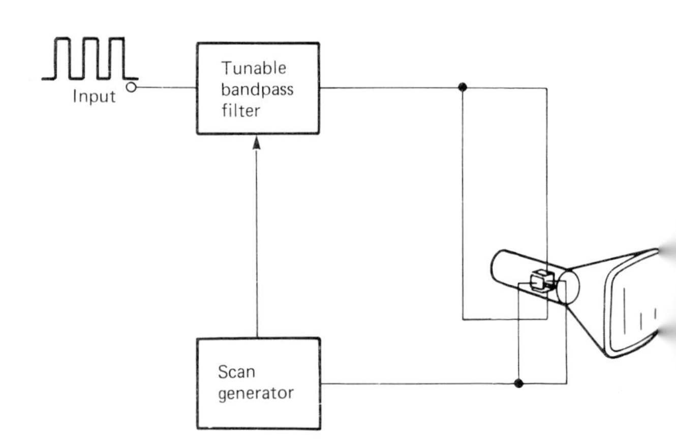

Figure 7.4 Block diagram of a swept tuned spectrum analyser

There are significant advantages in other ways, however. Only a single filter is used, so the overall circuit is much simpler and cheaper. Resolution bandwidth may be much improved, because the tunable filter may be designed to have only a small, well-defined bandwidth, although it is difficult to make such filters with sufficient accuracy and tuning range to realise a high performance spectrum analyser.

The third method used in spectrum analysers is the **swept superheterodyne spectrum analyser,** shown in block diagram form in Figure 7.5. The output of a voltage-controlled local oscillator is mixed with the incoming input signal to produce intermediate frequency signals, which are amplified by IF amplifiers. A scan generator provides the control voltage, sweeping the local oscillator frequency over the required range. This method has many advantages: high sensitivity due to the IF amplification; extremely wide frequency range (from only a few tens of kilohertz up to many gigahertz); variable resolution bandwidth; reasonable simplicity. These advantages are such that most modern spectrum analysers are of swept superheterodyne design. Like

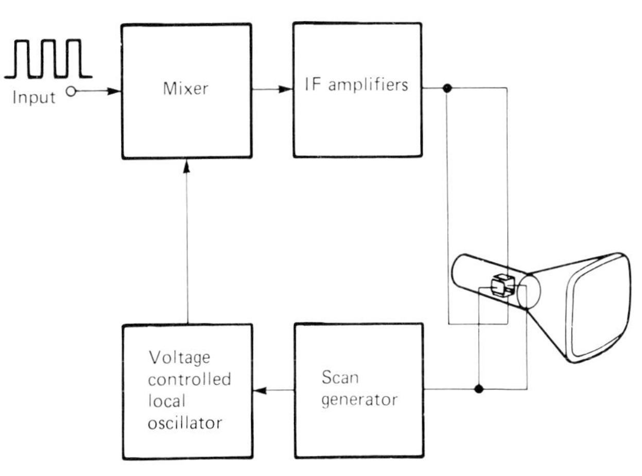

Figure 7.5 Block diagram of a swept superheterodyne spectrum analyser

the swept tuned spectrum analyser, however, the swept superheterodyne spectrum analyser is not a real-time test instrument.

All the methods so far discussed process the incoming input signal in an analog way, using filters or mixers to derive the frequency and amplitude information to be displayed. A fourth method of spectrum analysis, however, is one in which incoming signals are captured and *digitally* processed to derive frequency and amplitude information. The digital process is Fourier transformation, and so analysers using the method are called **Fourier transform spectrum analysers,** sometimes **fast Fourier transform (FFT) spectrum analysers.** Analysis takes place in close to real-time terms and FFT spectrum analysers are available that are capable of analysing signals up to about 100 MHz.

Photo 7.2 Rohde & Schwarz FSA spectrum analyser and display (Rohde & Schwarz)

Display characteristics

It is most usual to display signal amplitude in a logarithmic form, with one vertical graticule division corresponding to 10 dB difference in amplitude. Thus with a standard vertical graticule size of eight divisions (that is, the graticule is the same as an oscilloscope's – ten horizontal divisions by eight vertical) the spectrum analyser with this **display range** is capable of displaying amplitude differences of 80dB, a ratio of 10 000:1. The maximum display **dynamic range** that typical spectrum analysers can have is therefore 80 dB (dynamic range is the ratio of the largest to the smallest signals that can be displayed simultaneously), although a greater dynamic range range is possible with a larger graticule. Compared with the 40:1 range of the oscilloscope (that is, 32 dB), the spectrum analyser has a far greater range, and may be far more useful in certain comparison tests.

One problem with such a large dynamic range, on the other hand, is that two displayed signals which appear of a very similar size may, in reality, be quite different. For example, differences in size of only 0.15 divisions correspond to an actual difference of 3 dB, enough to define a system's bandwidth yet small enough so that it may not be detected on the display. So, spectrum analysers commonly have other display ranges, such as 2 dB per division, and a linear mode in which signals are displayed (as on the oscilloscope) with a calibration of volts per division.

Unlike the oscilloscope display, on which the centre horizontal graticule line is typically used as a reference, the spectrum analyser measures all displayed signals with reference to the top line. The user sets the front panel controls so that the maximum displayed signal on the display reaches the top line, and smaller signals are then measured as the number of divisions down from the top.

Generally, the central vertical line of the graticule corresponds to the central frequency of the swept range. Many spectrum analysers indicate this with a

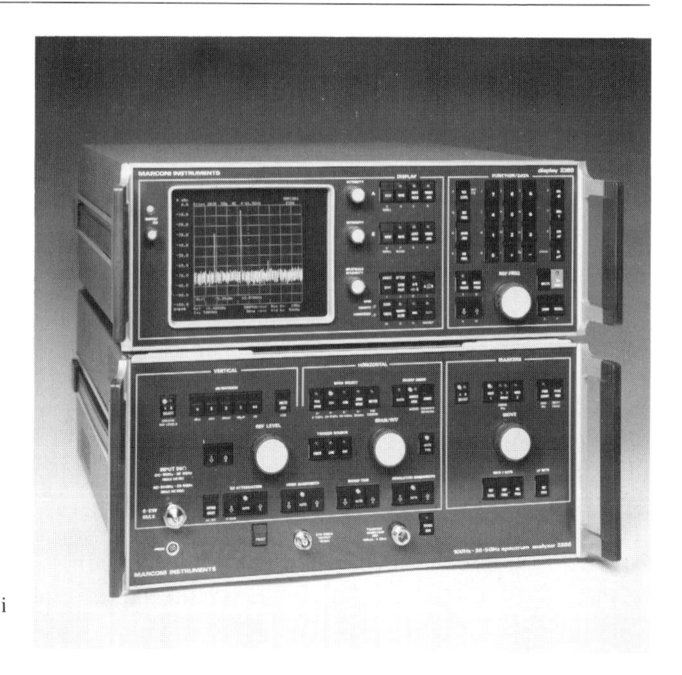

Photo 7.3 Marconi 2386 spectrum analyser and display (Marconi Instruments)

mark such as a dot. Front panel controls allow the user to set this central frequency and also the displayed frequency scan. Frequency scan is often defined on a 'per division' basis. A central frequency of, say, 100 kHz is first set, followed by the required frequency scan of, say, 10 kHz per division, corresponding to an observed range of 50 kHz to 150 kHz. Other frequency scan modes are on a 'full scan' basis, in which the total frequency range coverable is displayed, and 'zero scan' in which only the central frequency is displayed. In zero scan mode the spectrum analyser displays amplitude variation, at the central frequency, against a horizontal axis of time – so, in fact, it is a time domain display.

Figure 7.6a illustrates a possible spectrum analyser trace, which shows the presence of a signal component at the central frequency. At a level of about 60 dB below the maximum amplitude of the component (the display range is 10 dB per division) the **noise**

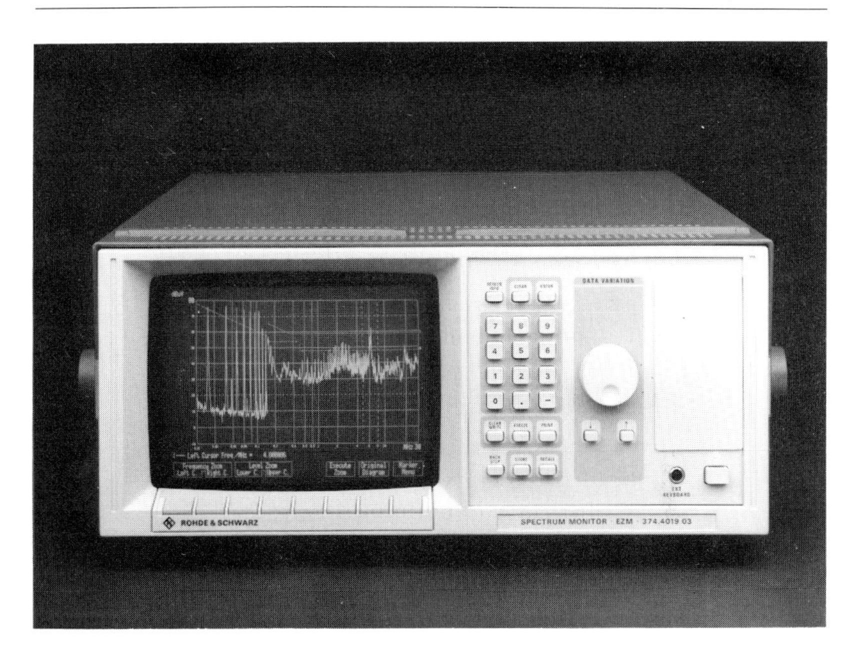

Photo 7.4 Rohde & Schwarz EZM spectrum monitor (Rohde & Schwarz)

floor, that is, the lowest horizontal part of the trace, exists. Because of its appearance, the noise floor is often nicknamed the **grass.** The resolution bandwidth of this display is the same as the frequency span per division setting.

Bearing in mind the earlier discussion about resolution bandwidth and the possibility of obscuring closely associated signal components. Figure 7.6b shows a possible trace obtained with a frequency resolution setting of about one-tenth the frequency per division setting. Two things are apparent: first, two signal components, closely situated around the central signal component, are now displayed; and second, the noise floor level has fallen by 10 dB (this is a general figure corresponding to a decrease in noise floor of 10 dB for each decade decrease in resolution bandwidth).

Decreasing the resolution bandwidth by a further decade, so that it is about one-hundredth the frequency span per division setting, reduces the noise

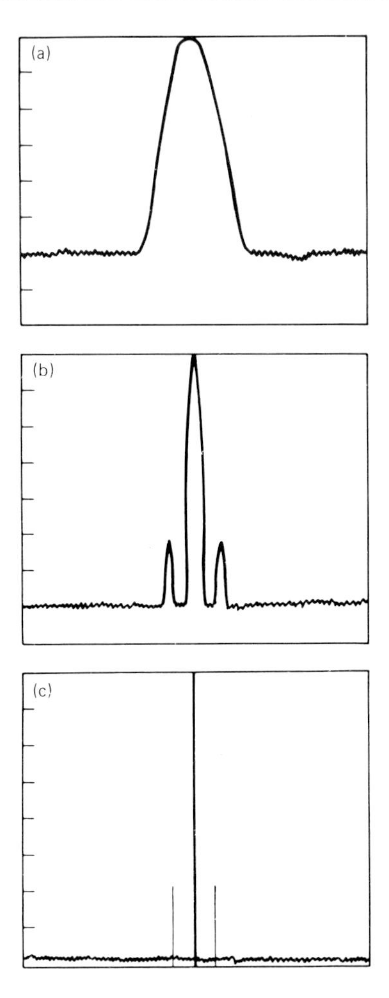

Figure 7.6
Illustrating the effects of varying a spectrum analyser's resolution bandwidth (a) with the resolution bandwidth set too wide (b) with the optimum resolution bandwidth (c) with too narrow a resolution bandwidth

floor by a further 10 dB and, one may expect, show up any more signal components closely situated around the central component. However, as Figure 7.6c illustrates, this is not always the case. As shown, the noise floor falls, but the signal components have become so thin as to be almost undetectable and there is a danger they may be overlooked. Going back to the earlier analogy of drawing lines with different thicknesses of pen tips, this is the equivalent of drawing lines with such fine pen tips that *they cannot be seen*. Furthermore, the smaller the resolution bandwidth, the slower the sweep time across the frequency range has to be, otherwise correct amplitudes of these reduced thickness signal components would not be displayed. There is, therefore, an optimum resolution bandwidth setting which depends on frequency span, sweep time, and signal amplitudes. Most modern spectrum analysers are capable of automatically optimising this, although users may still wish to check the settings manually, for accurate results.

Analyser sweep time is a variable control, determining the rate at which the analyser sweeps through the defined spectrum. If too fast, the analyser circuits may ring, or even fail to reach full amplitude. If too slow, the displayed spectrum is unobservable as a whole. Some spectrum analysers feature facilities to store and display the trace, however slowly it is formed, to allow exceedingly slow sweeps to be made of the spectrum. In modern analysers digital storage is used, whereby the display is divided into small horizontal segments and the amplitude of the trace in each segment is digitised and stored in RAM.

Equipment specifications and accuracy

There are four main classifications of spectrum analysers, which depend on the spectrum to be observed. The low frequency spectrum (below 100 kHz) is covered by real-time and FFT spectrum analysers. Above this, the radio frequency (100 kHz to 2 GHz), microwave (2 GHz to 21 GHz) and millimetre wave (above 21 GHz) spectrums are generally covered by swept superheterodyne spectrum analysers.

Main specifications of importance in spectrum analysers are frequency accuracy and stability, resolution bandwidth, and dynamic range. Of these, frequency accuracy and stability is most critical, and in modern swept superheterodyne analysers (the most common) is accomplished with phase locking techniques applied to three or even four stages of superheterodyne conversion. High-quality analysers of this type have a specification of stability around $\pm 10^{-6}$ per year, resolution bandwidth of about 1 in 10^8, and dynamic range of 100 dB.

Specifications of secondary importance are amplitude measurement range, amplitude accuracy and sweep time range. An amplitude measurement range of about −130 dBm to +30 dBm is typical, with high-quality equipment measuring from about −160 dBm to + 30 dBm. Accuracy over the range should be better than about ± 5 dB. Sweep time range depends upon requirements, but between 100 ms to 100 s (that is, 10 ms per division to 10 s per division) is common, with a range between about 1 μs to 1500 s (0.1 μs per division to 150 ms per division) being the maximum available.

8
Logic analysers

In Chapter 4 the need for oscilloscopes capable of storing and displaying singly occurring, that is, non-real-time information, was discussed. Typically, this sort of information occurs within digital systems. The digital storage oscilloscope was suggested as a test instrument which may be used to display non-real-time information, and in many simple situations will perform adequately. However, in more complex situations, say observation of the information flow on a microprocessor system's data bus or address bus, the digital storage oscilloscope with its two or

Photo 8.1 Rohde & Schwarz LAS logic analysis system (Rohde & Schwarz)

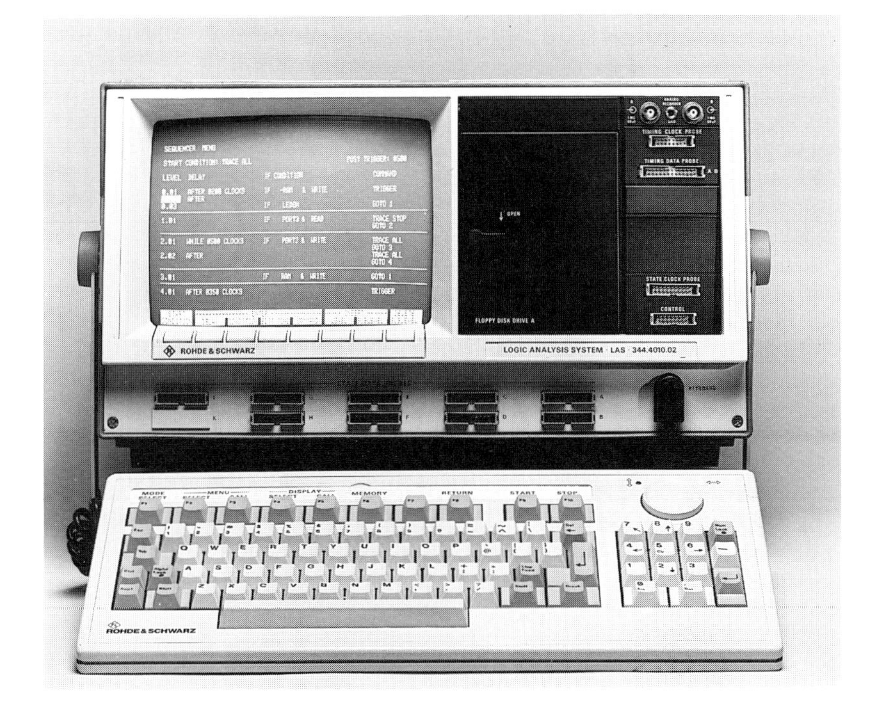

even four inputs cannot possibly display sufficient simultaneous information, purely because so many signal lines exist. Compounding the problem is the fact that signals in digital systems cannot be monitored in isolation, as they only have significance when observed along with all other signals in the system. In other words all relevant signals in digital systems have to be monitored together, and the digital storage oscilloscope as it stands is not capable of this.

The **logic analyser** is a development of the digital storage oscilloscope and performs similar non-real-time display functions. Instead of a maximum of four signal inputs, however, the logic analyser has 8, 16, 32, 48 or more signal inputs, depending on complexity and, of course, cost. Here, however, the similarity between digital storage oscilloscopes and logic analysers ends. The oscilloscope is a hardware-based test instrument built for the specific purpose of displaying signals in time domain form, whereas the logic analyser is microprocessor-controlled and software-based, with many more additional features than just a simple time domain display of the stored signals. It can manipulate the stored signals, reformatting them into alphanumeric data for display and can even present the data in a way which shows how the observed microprocessor system's software actually runs, step by step.

As the logic analyser *is* a microprocessor-based system in its own right, that allows us to consider the device internally. Figure 8.1 shows in block diagram form a logic analyser with all the usual microprocessor system parts we would expect: microprocessor (consisting of control unit, arithmetic logic unit, and registers), ROM, RAM, inputs and outputs, and buses.

Being microprocessor-based, most logic analysers are general-purpose; not only in that the functions

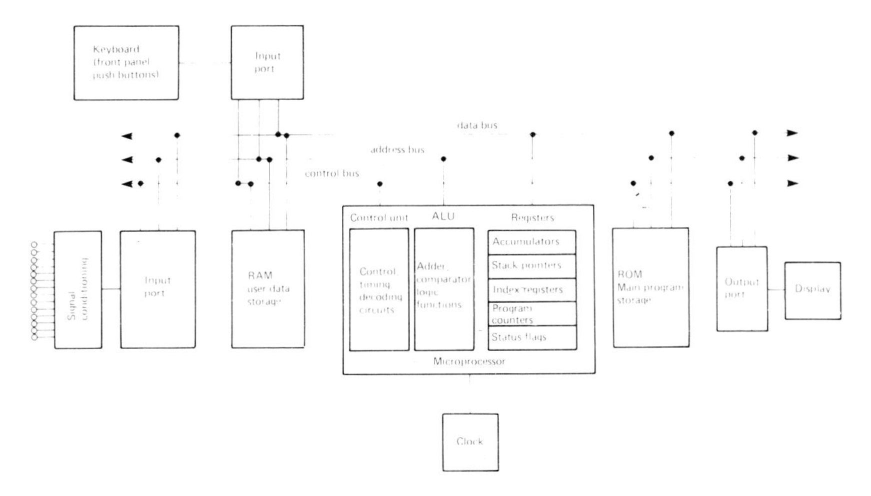

Figure 8.1 Block diagram of a logic analyser

they perform are typical of all logic analysers, but in that they may be used to measure signals from all types of microprocessor systems. Connecting a general-purpose logic analyser to a specific microprocessor system is a simple matter of interfacing the two with a special-purpose interface module, which synchronises the logic analyser to the microprocessor system, formats data from the microprocessor system in a way the analyser can understand and provides the necessary electrical and mechanical connections. Thus, say, one interface module may be used to connect the logic analyser to any Z80-based microprocessor system, while another module would be used to connect the analyser to any 8085-based microprocessor system, etc., etc. Different manufacturers have different names for interface modules, such as personality option devices or preprocessors, but they are most commonly nicknamed **pods.** Usually pods are separate devices, linked to the logic analyser and microprocessor system with ribbon cables and plug-in connectors, but sometimes, particularly on specialised logic analysers, the pod functions are performed by printed circuits which plug in to the analyser. When coupled with a pod

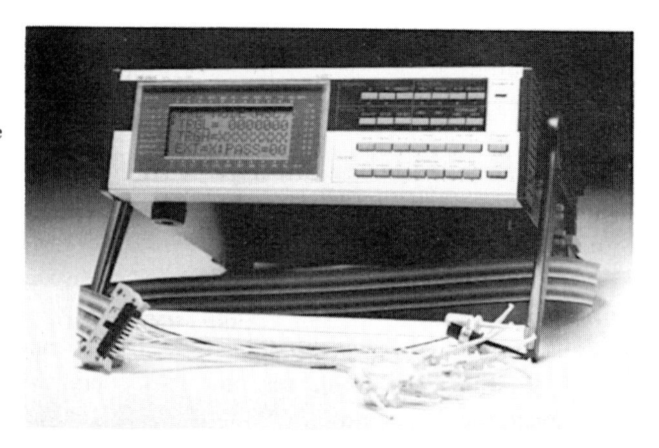

Photo 8.2 Soar 1420 logic analyser with liquid crystal display (Advance Instruments)

to a microprocessor system the general-purpose analyser becomes a **composite logic analyser,** more commonly known as a **microprocessor-specific logic analyser.** Some logic analysers are manufactured as microprocessor-specific test instruments, but this obviously limits their use to only one type of microprocessor-based system, and so such analysers tend to be very specialised and very expensive.

Features

The number of input channels a logic analyser has more-or-less defines the instrument's potential. If the logic analyser is to be used to analyse an 8-bit microprocessor system with, say, a 16-bit address bus, an 8-bit data bus, and five or six control lines, the minimum number of input channels the analyser ideally needs is about 30. To analyse fully a 16-bit microprocessor system similarly requires a logic analyser with about 46 or so input channels. Input impedances are similar to those of oscilloscopes, that is, typically around 1 MΩ in parallel with about 10 to 15 pF.

Another factor determining the logic analyser's usefulness is its triggering flexibility. Unlike the oscilloscope, which is usually triggered by the input signal's first crossing of a preset threshold, the logic analyser must have many other optional trigger modes. Some important ones are:

- triggering on occurrence of a predetermined data word; that is, when the input signals form a particular data combination, the analyser is triggered
- pre/post triggering, to store and if necessary display input data before and after the trigger word
- triggering when a 'glitch', that is, a discrepancy between the expected signals and those obtained, occurs in the input data
- storing and displaying the data present at a preset interval (usually a number of clock cycles) *after* the trigger word
- storing and displaying the data present after a specified number of occurrences of the trigger word
- comparing actual data events with expected events, and triggering only when they do not coincide.

Memory size and speed of memory operation are important. Memory is often sectioned-up into various segments to store different types of data such as reference data, high-speed signal data, and stored signal data, with the user defining which memory segment is used for which function. Typical segment size is between about 64 words to 2K words, and typical number of segments is 16. The **time window,** that is, the length of time over which signal data can be observed, is directly relevant to the logic analyser's total memory size. In situations where not all the logic analyser's signal input channels are required, it is often possible to couple memory segments into consecutive links, so that the fewer

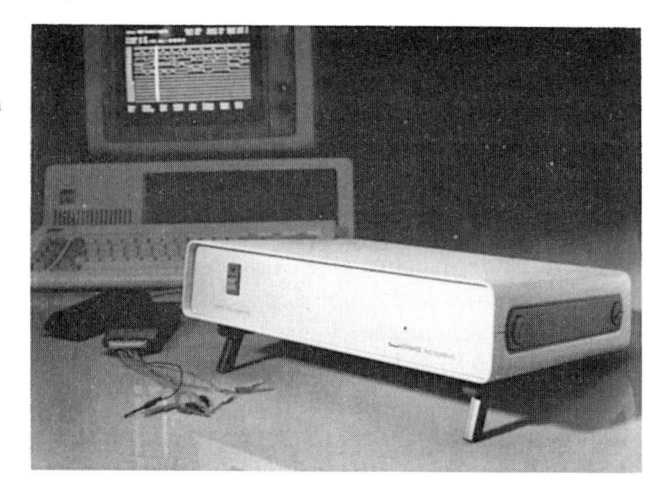

Photo 8.3 Advance 1950 modular logic analyser, shown in use with personal computer (Advance Instruments)

signal inputs can be viewed over a longer time window.

Other features which a logic analyser may have are more varied. Often a digital word generator is in-built, which provides a stimulus to the microprocessor system under observation. The logic analyser may then be used to observe what happens upon this stimulus. More complex logic analysers often have non-volatile storage (EEPROM, battery-backed CMOS RAM, or similar), to read and write previous data inputs, set-up information, or reference information to compare with incoming information. Sometimes, floppy disk drives are used for larger data storage requirements. Many logic analysers allow direct comparisons to be made, by highlighting differences, between newly stored data and reference data. Sometimes logic analysers are provided with two or even three system clocks to allow data input from, say, multiplexed microprocessor systems.

This range of features is not a fixed one and largely depends on user-requirements. Also, because the logic analyser is itself a microprocessor-based test instrument, features can sometimes be changed by

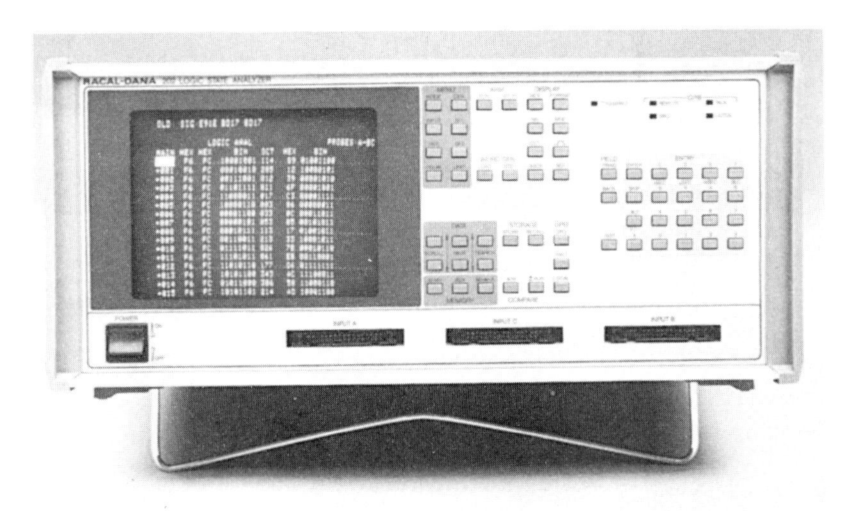

Photo 8.4 Racal-Dana 202 logic state analyser (Racal Group Services)

manufacturers of logic analysers merely by incorporating software updates.

Main modes of use

Logic analysers have developed from digital storage oscilloscopes and, like them, generally have cathode ray tube displays, although a small number of recent analysers have a flat-panel liquid crystal display. One of the logic analyser's main functions, also like the digital storage oscilloscope, is to display signal inputs as functions of time. This mode of operation is known as **timing analysis** and is used to observe signals immediately before and after a particular trigger point. Due to the limited size of display used in logic analysers, generally only eight of these input channels are displayed at any one time.

Samples of signal levels at each input channel are taken at every clock cycle. However, these samples are not analog samples as, being a *logic* analyser, the device assumes the signals are at one of two logic levels. Waveforms displayed are therefore not amplitude against time graphs, as in the oscilloscope, but logic levels against time graphs.

A movable cursor is generally available, which may be moved over the display. Often the word value at the cursor position and time interval or number of clock cycles between trigger and cursor are displayed in alpha-numeric form, allowing accurate time measurements to be made.

As this description of sampling at every clock cycle suggests, timing analysis requires asynchronous operation of the logic analyser and so the analyser's own internal clock controls the procedure. In this respect timing analysis *may* be undertaken without the use of an interface pod between analyser and system, although a pod is still preferred to aid electrical, mechanical and format interfacing. This asynchronous operation tends to use up quite a lot of memory as, to ensure that each change of logic state is sampled, samples must be taken at least as often as any expected logic state change. Now, the advantage of using timing analysis is that the user can view any short discrepancies or glitches which may upset the observed microprocessor system's operation. Such glitches may be of only a few nanoseconds length, consequently the logic analyser samples must be taken in the same order of time. In practice the shortest pulsed logic change (either true pulse or a glitch) that can be typically displayed by a logic analyser is about 1 to 1½ times the internal clock cycle period and is referred to as the logic analyser's **resolution.** Resolution is probably the most important aspect about a logic analyser and, as can be seen, is primarily governed by the internal clock frequency: the faster the clock, the more frequently sampling can take place. A typical high-resolution logic analyser is able to capture and display glitches and pulsed data changes of less than 10 ns length, corresponding to a clock frequency of about 100 MHz.

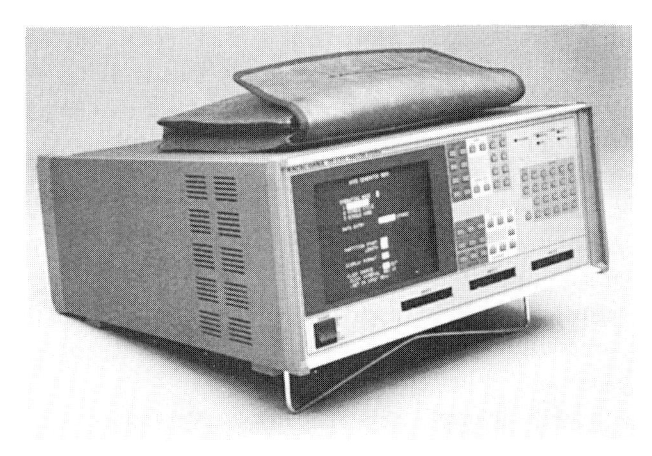

Photo 8.5 Racal-Dana 205 logic state analysis system (Racal Group Services)

Some specialised analysers have a clock frequency of over 600 MHz, so they can capture and display glitches of around 2 ns. Even if a logic analyser cannot display a glitch (because analyser resolution is not high enough) it may still be able to capture it, and the user may be aware of the glitch's existence because the display may signify a changed and thus unusual, logic state. Manufacturers will often quote the analyser's glitch capture capability as being around three or four times its actual resolution, for this reason. Whatever the logic analyser's resolution, however, the reader will appreciate that timing analysis is used to pinpoint logic performance in the observed microprocessor system over very small periods of time.

On the other hand, the other main mode of logic analyser use is to check *overall* logic performance and so synchronous operation, where the logic analyser uses the microprocessor system's clock for timing purposes, is preferred. Thus, logic state samples are taken only at every cycle, or multiple, of the microprocessor system's clock and more of the complete operation is stored within the logic analyser's memory. This **state analysis** of input signals presents the logic state information in binary form, as a collection of ones and zeros grouped into words on the display, as it is stored in the analyser's memory.

One of the functions that a pod helps the logic analyser to perform is to convert the parts of this information corresponding to the machine code into the assembly mnemonics of the microprocessor system under observation. This **reverse assembly** function of the logic analyser, commonly called **disassembly,** is really just a sub-mode of state analysis and can be most useful, as it allows the microprocessor system engineer to observe internal operations directly in terms of easily recognised and already familiar mnemonics, rather than difficult to understand machine code. Other options of state analysis performed by pods allow the binary information to be displayed as octal, hex, decimal, ASCII, EBCDIC, etc. formats.

In simple terms, checking a microprocessor system's operation will usually entail a combination of state analysis and timing analysis. For example, the engineer may use state analysis to check overall software performance; that is, make sure the program is running correctly. If it isn't, state analysis will show the incorrect area. Next step is to observe the faulty area, with timing analysis. Drawing an analogy, it's rather like taking a photograph using a zoom lens. First, you pan across the overall scene at least magnification until the object to be photographed comes into view. Then, you zoom in on the object and shoot. For this reason, some logic analysers allow a great deal of state and timing analysis **interaction** between the two modes. A useful feature, for example, on some analysers allows time measurements to be recorded and displayed at intervals when in state analysis. Thus the area to 'zoom in' on when resorting to timing analysis is apparent.

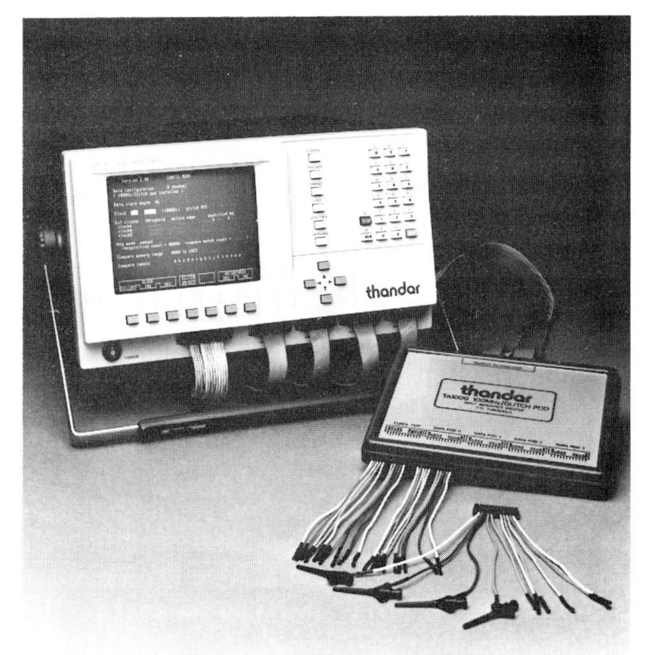

Photo 8.6
Thandar TA1000 logic analyser with glitch pod (Thurlby-Thandar)

A third use of logic analysers has arisen as they have developed, and is known as **performance analysis.** In this mode the microprocessor system's general software performance is monitored by the logic analyser. Displays in the form of graphs and histograms may show such factors as system address usage, program statement execution times, etc. Additional modes of use, such as performance analysis, are a feature of logic analysers because, being microprocessor-based test equipment, they will constantly be developed in terms of software as users demand more functions from the equipment manufacturers. New modes of use will evolve over the coming years.

9 Time domain reflectometers

Borne of the non-real-time storage oscilloscope principle, in which a single occurring measurement is to be displayed, the **time domain reflectometer** (TDR) is used to test and measure parameters in communications networks, with the specific purpose of identifying and, more important, locating faults.

In simple terms, it does this by transmitting a defined signal into the network and displaying the returned signal. Figure 9.1 illustrates the principle where a signal is transmitted and received by the equipment. A signal (in this case a single pulse as shown in Figure 9.1a) is initially transmitted, and after a time is returned in a form adapted by the network (Figure 9.1b) to be displayed (Figure 9.1c).

Form the signal takes as it is returned to the time domain reflectometer is unique to the point at which it is transmitted. For this reason the shape of the display is sometimes said to be a **signature**. The fact that similar faults in networks produce similar signatures means users can identify faults. Also, as the signature display is in the time domain (that is, the display's horizontal axis is a function of time), time the signal takes to be transmitted between the time domain reflectometer and the fault is measured – so the distance between is found simply as the reciprocal.

There are two main categories of time domain reflectometer:

- **metallic time domain reflectometers** (MTDRs – used in metallic cable communications systems; co-axial, twisted pair and so on

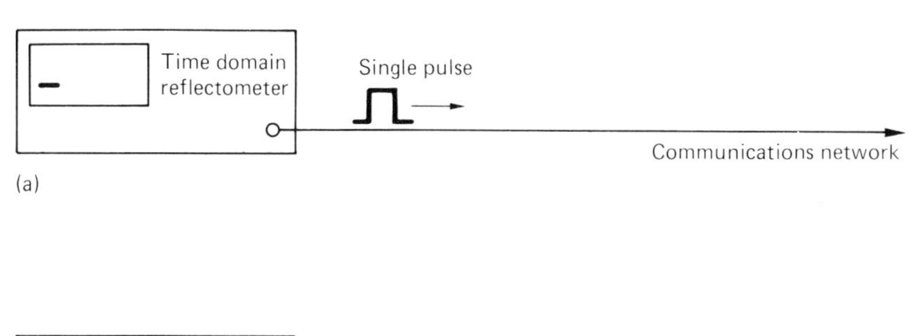

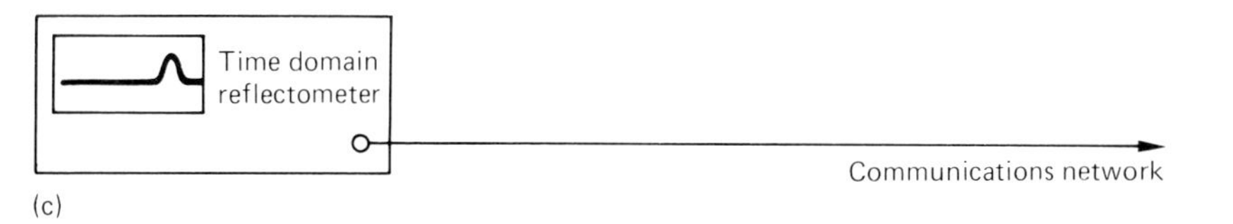

Figure 9.1 Principle of transmission and return of signal (a) initial transmission (b) returning (c) display

- **optical time domain reflectometers** (OTDRs) – used in optical cable communications systems; usually optical fibre networks, giving rise to the alternative name *fibre optical time domain reflectometers* (FOTDRs).

Considerably different techniques are used in each category, so they must be discussed separately.

Metallic time domain reflectometers

Time domain reflectometers of this category are a fairly straightforward adaptation of the digital storage oscilloscope principle. Figure 9.2 shows a block diagram of a typical metallic time domain reflectometer, where the main parts of an oscilloscope may be seen, together with a few new blocks.

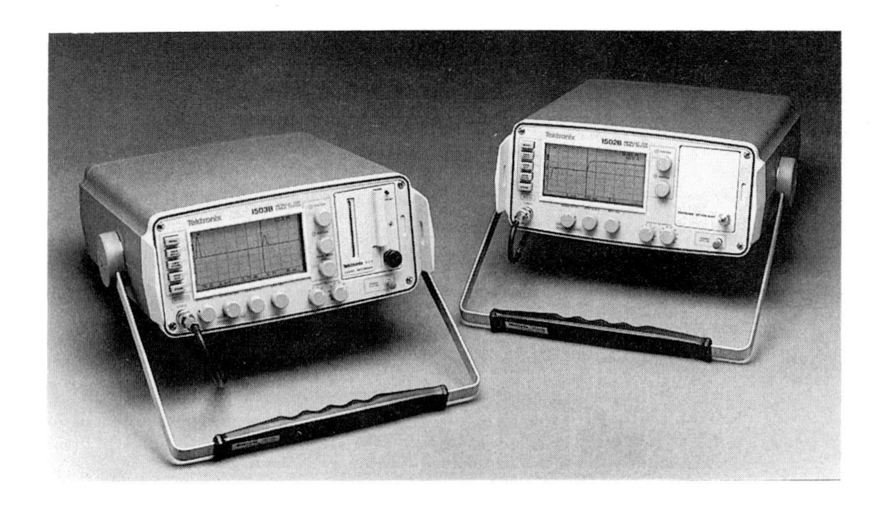

Photo 9.1 Tektronix 1502B and 1503B metallic time domain reflectometers (Tektronix)

A signal generator produces a test signal every time it is triggered by the noise filter trigger. The noise filter trigger operates on the principle that noise is reduced by signal averaging, that is, if a number of received signals are averaged prior to display the signal is strengthened: random noise received along with the signal, on the other hand, will be reduced in comparison with the strengthened signal. Noise filter triggers with variable averaging from around 1 to 128 cycles are common.

Received signal is initially amplified in a vertical pre-amplifier. Averaging of the signal then takes places in a digital averaging store, which is triggered by the noise filter trigger every time the required number of averaged signals has been received. The averaged signal is then passed on to the vertical amplifier, which features a gain control (labelled vertical scale on the front panel) and a Y shift control.

At the same time as triggering the averaging store, the noise filter trigger triggers the horizontal time-base. Timebase has speed control, generally labelled as a distance per division control on the front panel.

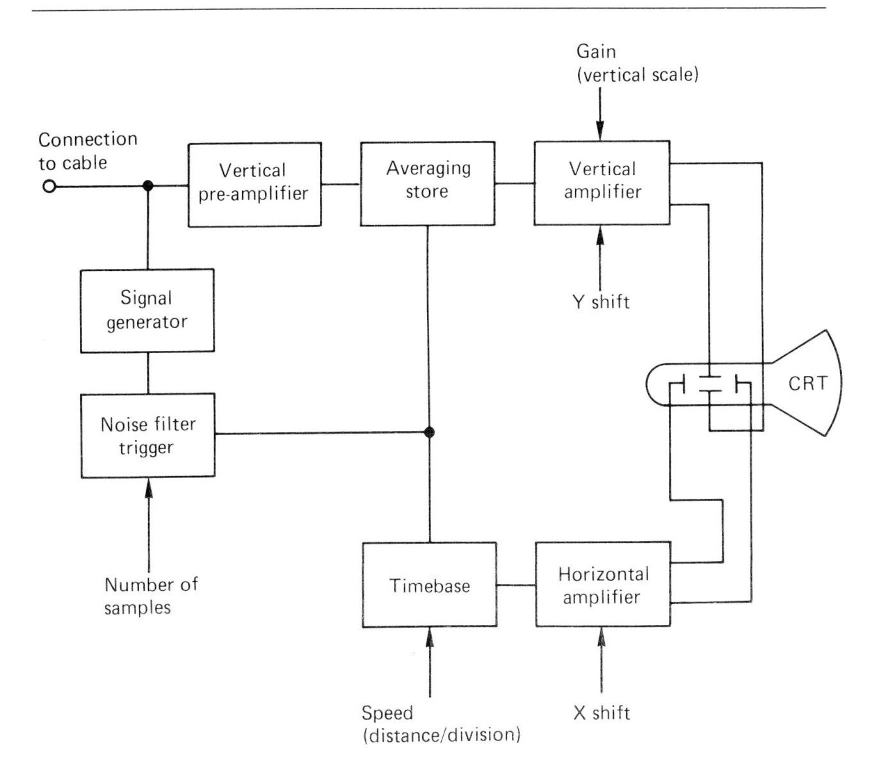

Figure 9.2 Block diagram of main parts of a metallic time domain reflectometer

Timebase signal is then passed to the horizontal amplifier, which features an X shift control.

Although Figure 9.2 shows a cathode ray tube as the display device, it is becoming common for time domain reflectometers to feature a solid-state display, such as an LCD screen. These have the advantage of low weight, small size and battery-powered portability where required.

Fault detection

As the network is metallic, whether co-axial, twisted pair, or multistrand, it acts as a transmission line. So, when no faults are present the transmitted signal is totally absorbed into the load, and no signal is returned back to the metallic time domain reflectometer. However, if any fault exists, part or all the

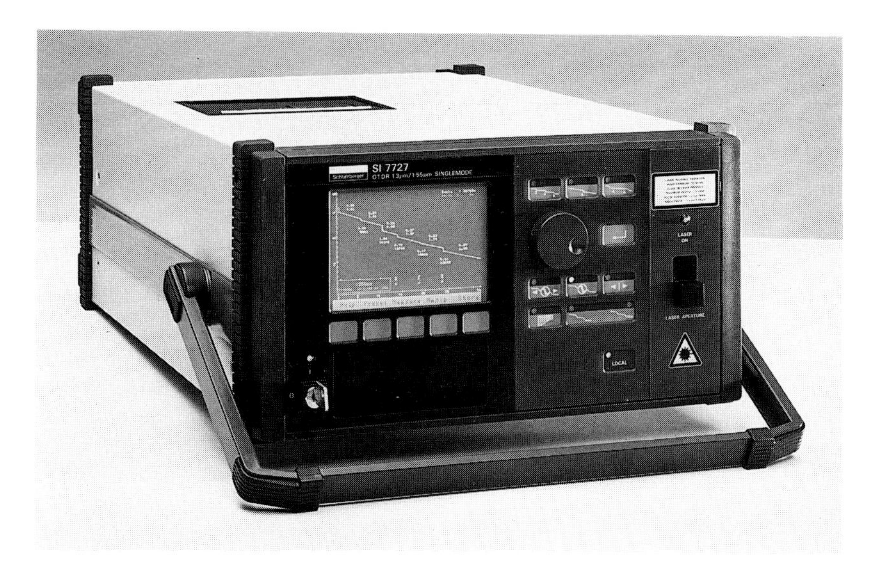

Photo 9.2 Solartron SI7727 optical time domain reflectometer (Schlumberger)

signal is reflected back towards the time domain reflectometer from the fault. If more than one fault is present, more than one reflection occurs. In this light, metallic time domain reflectometers can be understood to operate in the same way as radar systems.

Size and polarity of the reflected signal depend on the type of fault. At the extremes, a short circuit produces a total reflection of the transmitted signal with complete polarity reversal. An open circuit, on the other hand, produces a total reversal with no change in polarity.

Generally, the reflection coefficient ρ (represented by the Greek letter *rho*) is used to relate reflected signal to transmitted signal according to the expression:

$$\rho = \frac{\text{reflected signal amplitude}}{\text{transmitted signal amplitude}}$$

such that total reflection with no polarity reversal is represented when $\rho = +1$, and total reflection with polarity reversal is represented when $\rho = -1$.

Reflection coefficient within these limits means that reflection is not total. If ρ is positive the overall cable's impedance is above the characteristic impedance; if ρ is negative the overall cable's impedance is below the characteristic impedance. If ρ is zero, the cable impedance is equal to the characteristic impedance, that is the cable is correctly terminated with no faults.

Metallic time domain reflectometers vary in two main operational ways:

- **signal type** – there are two basic signal types used by metallic time domain reflectometers: a stepped change in applied voltage; a single voltage pulse. Figure 9.3 illustrates the principle behind step-

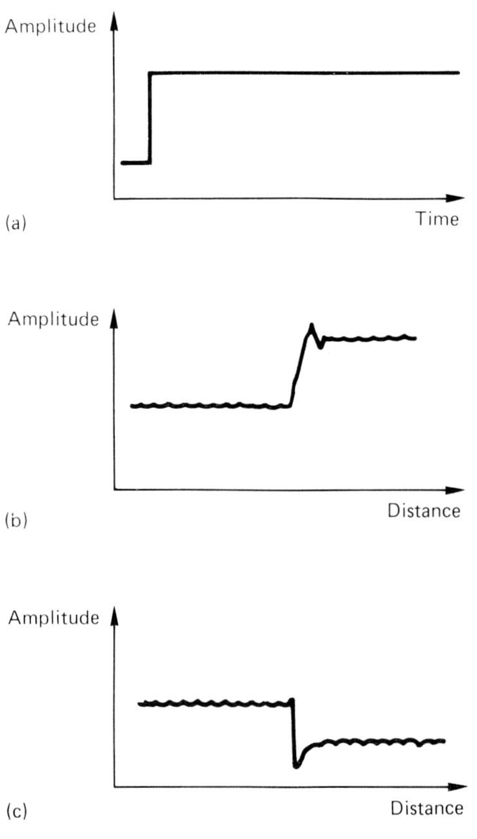

Figure 9.3 Principle of stepped change metallic time domain reflectometers (a) transmission of stepped change (b) a step-up reflection indicates an inductive or higher resistance fault (c) a step-down reflection indicates a capacitive or lower resistance fault

ped change metallic time domain reflectometers, where a stepped change in voltage with a fast rise time is transmitted along the cable under test (Figure 9.3a). The reflected voltage is therefore also a stepped voltage, with the effects of the reflection superimposed on the step. Faults that are inductive or of higher resistance than the cable impedance cause the reflection to be a step-up transition (Figure 9.3b), while faults which are capacitive or of lower resistance than the cable impedance cause the reflection to be a step-down transition (Figure 9.3c). The height of the reflection is a measure of the impedance change at the fault.

Figure 9.4 illustrates the principle behind pulsed metallic time domain reflectometers. Here a single well-defined pulse (usually one half of a sine cycle) is transmitted along the cable under test (Figure 9.4a). Overall shape of the reflected pulse is therefore also of a half sine cycle. Polarity, just as the reflected signal's polarity in the stepped change metallic time domain reflectometer, depends on the fault (Figure 9.4b and c) and height, too, depends on the impedance change at the fault

- display type – often metallic time domain reflectometers display the returned signal directly in terms of the reflection coefficient ρ as the vertical axis, and distance as the horizontal axis. It is a straightforward job of reading the height of the returned signal at a specific distance along the horizontal axis to determine the impedance of the fault, and the distance along the cable.

Some metallic time domain reflectometers, on the other hand, use a different scale for the vertical axis. One example is **return loss**, which is related to ρ according to the expression; return loss $= 20 \log_{10}\rho$.

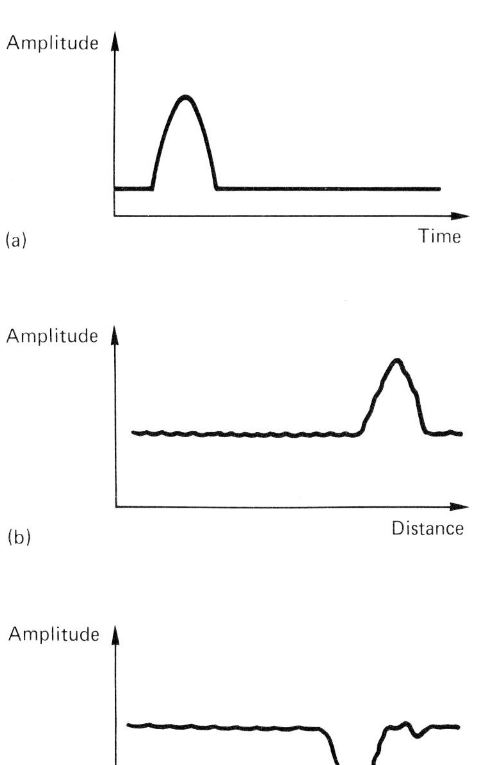

Figure 9.4 Principle of pulsed metallic time domain reflectometers (a) transmission of pulse (b) reflection of the same polarity indicates an inductive or higher resistance fault (c) reflection of the opposite polarity indicates a capacitive or lower resistance fault

This is a logarithmic expression, measured in decibels. At the same time, the equipment now makes use of an internal logarithmic amplifier to make relatively greater amplification of faint signals possible. Such a scale and amplifier makes location of a specific fault easier, in that displayed signals are all now about the same size, although it makes measurement of the exact impedance of the fault more difficult. However, as metallic time domain reflectometers are normally going to be used merely to locate faults anyway, the actual measurement of the fault impedance is of relatively little significance. In effect, if the equipment is to be used as a fault finder, calibration in terms of a quantity such as return loss and a logarithmic amplification of received signal give the best scale. On the other hand, if it is to be used to predict exact quantities, a ρ scale is best. Some metallic time domain reflectometers feature scales switchable between the two.

Typically, vertical gain is variable between about 0.5 mρ per division and 500 mρ per division for equipment with a ρ scale. Return loss scaled equipment, on the other hand, will have a variable gain from about 0 dB to 60 dB.

Using the metallic time domain reflectometer

One of the most important points to remember when operating a metallic time domain reflectometer is the fact that cable loss inherent in the tested network causes the signal to be attenuated as it passes down the cable and back. For this reason, major faults at distant locations will give received signal reflections which appear no larger than minor faults close to the test equipment.

This makes little or no difference in the location and identification of faults, but it does make it difficult to measure absolute reflection coefficient at a point some distance from the test equipment.

Metallic time domain reflectometers are used in two main ways: to locate and identify a fault which is known to exist, and to ensure that new connections into a network are correctly performed.

Fault location

Location of a fault relies first on the accuracy of the horizontal scale displayed on the equipment. Generally this accuracy will be in the region of about ± 2%. How this scale relates to the cable being tested, however, is just as important. Users must remember that any metallic time domain reflectometer displays only the *electrical* length from the equipment to the fault with the stated accuracy. The *physical* length will be different, as it depends on the cable itself. Factors such as variations in cable propagation velocity, possible loops or snakes in the network (known as **snaking loss**), and even the fact that different types of cables may have been used must all be taken into account. The test equipment cannot be blamed for inaccuracies due to such factors. On a small cable network this may not prove critical, anyway, as a difference in measured electrical length to a fault and the actual physical location of that fault may only be a short distance anyway. However, on a large network, faults may be located electrically within just a couple of metres, say, but physically may not be locatable within an order of magnitude greater than this.

There are three ways in which fault location under such conditions may be aided:

- **network knowledge** – by using known points on the cable, the display may be accurately calibrated. For example, if it is known that a connection exists precisely 1000 metres from the test location, the front panel controls of the metallic time domain reflectometer allow the user to adjust the signature precisely to match
- **successive approximation** – if a fault is located as being electrically, say, 2000 metres away, but physical accuracy of no better than ± 5% (that is, ± 100 metres) can be expected, a closer reading (perhaps at a connection point 1500 metres along the cable) will give a distance of only 500 metres which at a physical acuracy of ± 5% is equivalent to ± 25 metres. Then, a reading somewhere within this 25 metre of cable will give a fault location with an accuracy of at most ± 1.25% – probably much better than this as the cable at the location

will be measured more easily and so inaccuracies due to the difference between electrical and physical lengths will be greatly reduced

- **measurement at either end** – allows a simple recalculation of location by a straightforward splitting of the difference, or a more accurate recalculation if the exact length of cable l is known, according to the expression:

$$d_a = \frac{l}{\left(\frac{d_1}{d_2} + 1\right)}$$

where d_1 is the measured location from end 1, d_2 is the measured location from end 2, d_a is the actual fault location from end 1.

Fault identification

Identification of a fault relies on the fact that the operator has a working knowledge of metallic time domain reflectometer signatures. This may be as simple as carrying around textbook examples, to look at after measurements of any particular cable network have been performed. Figure 9.5a to h shows some examples.

Figure 9.5 Textbook signatures obtained with various faults by metallic time domain reflectometers of both signal types: (a) and (b) inductive fault; (c) and (d) capacitive fault; (e) and (f) resistive fault; (g) and (h) connection to an incorrect impedance line – length of line of wrong impedance is measured on the signature

Network monitoring

Where a cable network is to be adapted in some way, measurement with a metallic time domain reflectometer immediately prior to and after the adaptation allows the quality of the adaptation to be measured. For example, say a metallic time domain reflectometer with a return loss display is located at a test point. A break is now made at the required point of adaptation (say an extra connection is to be made) and the equipment is used to measure return loss r_1. After the connection is made, the return loss r_2 is measured. The actual return loss r_a can now be calculated from the expression:

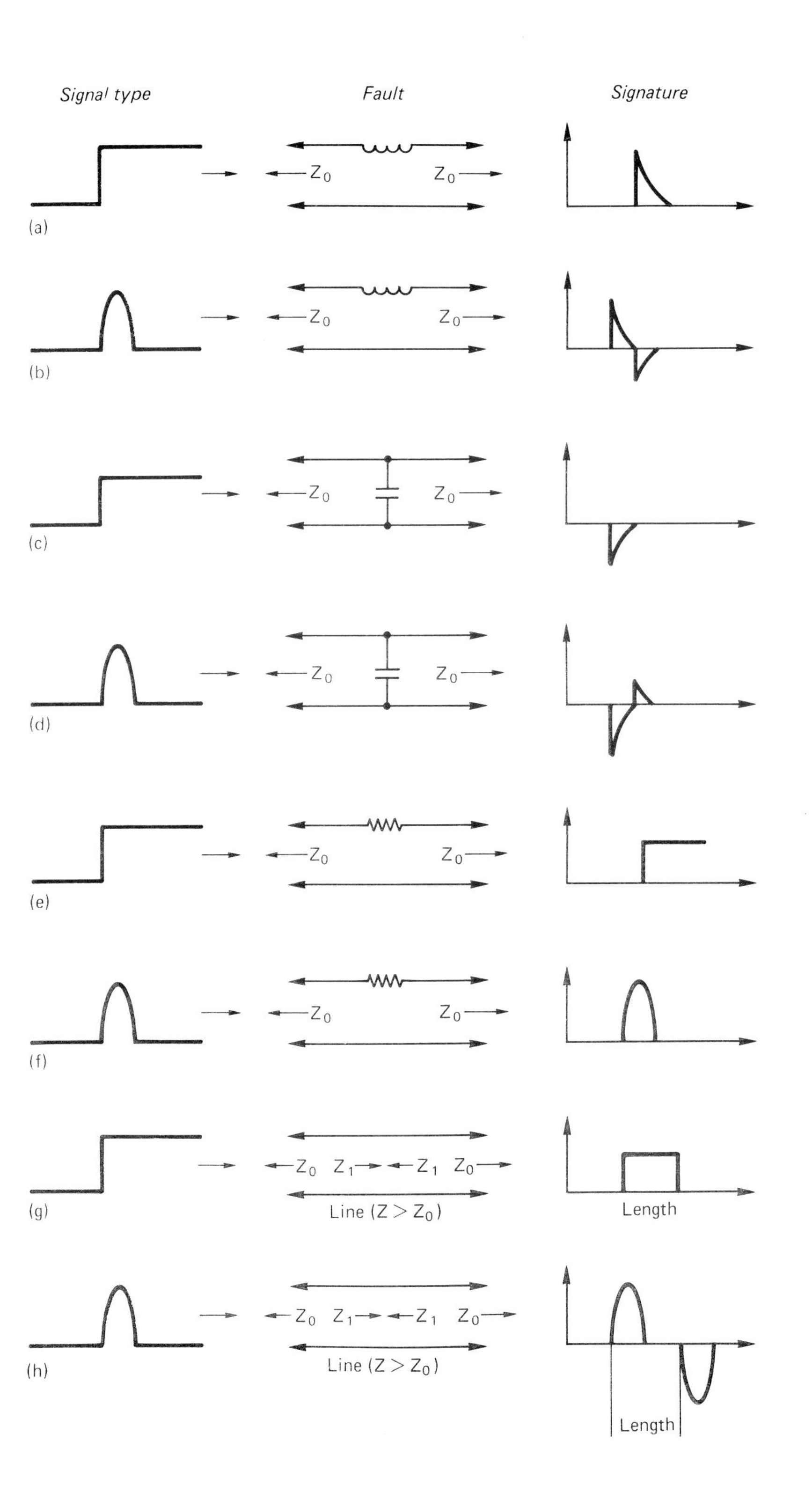
Signal type
Fault
Signature
(a)
(b)
(c)
(d)
(e)
(f)
(g)
(h)
Z0
Z1
Line (Z > Z0)
Length

$$r_a = r_2 - r_1 \text{ dB}$$

So, the reflection coefficient at the connection can now be calculated from the expression given on page 124, return loss $r = 20 \log_{10} \rho$, such that:

$$\rho = 10^{-(r/20)}$$

and impedance at the connection point is given by the expression:

$$Z = Z_o\left(\frac{1+\rho}{1-\rho}\right)$$

Optional features

A number of options are usually available for modern metallic time domain reflectometers. First, as measurements are momentary and it is often a useful feature to maintain a record of signatures, miniature dot-matrix chart recorders built into the equipment are often available.

Where adjustable controls are not provided to allow a range of cables with differing impedances, it is usual to be able to plug-in impedance matching adaptors.

Certain types of networks, particularly local area networks (LANs) such as the Ethernet LAN have quite specific signals present on the cable. For such measurement, manufacturers often provide adaptors which allow the metallic time domain reflectometer to be easily interfaced on to the network and check for faults using the network signals.

Optical time domain reflectometers

Where the metallic time domain reflectometer transmits a signal on to the cable under test and measures the signal returning after reflection, the optical time domain reflectometer transmits a pulse of light into a fibre and measures the signal returning after backscattering, a sort of refraction. Backscattering is simply light returned to the source end of a fibre due to Rayleigh scattering. Scattering is the result of light energy being redirected by the molecular structure of the glass which makes up the fibre.

Generally, the power of the backscattered light is some 45 dB below that of the forward light at any point on the fibre, although this depends on light wavelength and fibre type. For this reason, considerable gain has to be introduced by the optical time domain reflectometer on measurement. Significant signal averaging is also usually incorporated.

Power of the backscattered light, as a function of time after the source pulse, is a direct effect of the fibre itself and is used in the optical time domain reflectometer to give a display similar to that in the metallic time domain reflectometer, showing inconsistencies, faults and distances from the equipment. Unlike the horizontal display of the metallic time domain reflectometer (which allows a dual-polarity voltage to be displayed), however, the optical time domain reflectometer's display is a slope which shows power of the returned signal. As the power diminishes, the trace gets lower on the screen, so the slope is a more-or-less continuous line decreasing from left to right, as shown in Figure 9.6, with inconsistencies such as faults superimposed. Overall gradient of the slope is a measure of attenuation although, by nature of the measurement, is not exact.

Attenuation can be calculated from an optical signature over any portion of the slope. For example, if the attenuation caused by the connection shown in Figure 9.6 is to be measured, the power immediately prior to (P_1) and after (P_2) the connection may be estimated off the screen. The attenuation due to the connection A_C is then calculated from the expression:

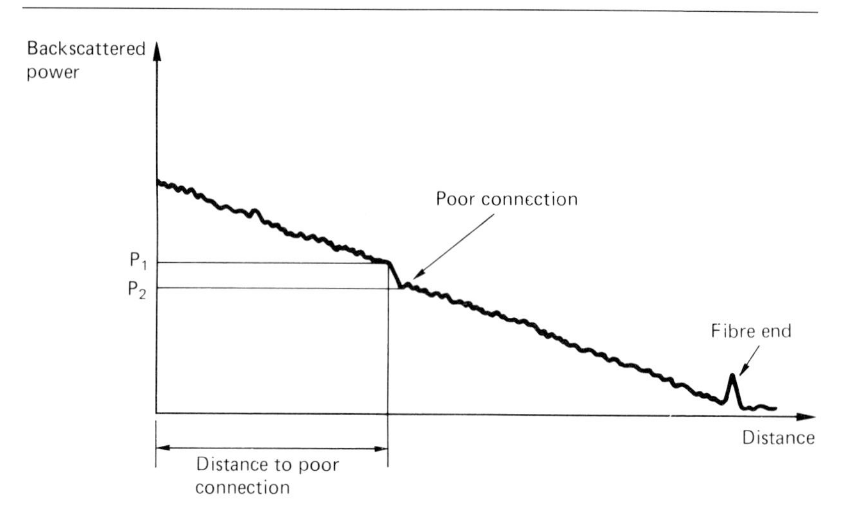

Figure 9.6
Typical signature from an optical time domain reflectometer with a fault caused by a poor connection

$$A_c = 5 \log_{10} (P_1/P_2)$$

This is, perhaps, a different expression to that which might be expected. Normally, power calculations in decibels use the factor 10, not 5. Reason is that the display shows *two-way attenuation*, that is, attenuation as the light passes through the connection from equipment outwards, plus attenuation as the light returns through the connection inwards. *One-way attenuation* is, of course, only half of this, in decibels. Modern optical time domain reflectometers sometimes feature an LCD readout which allows the equipment to calculate one-way attenuation and display the result. Users merely have to mark the two points over which attenuation is to be measured.

Generally, returned optical power is displayed on the screen in a logarithmic nature in switched or variable calibrated steps (from around 2 dB per division to 10 dB per division). Horizontal calibration is variable over a large range.

Optional features

Similar features to metallic time domain reflectometers are usually available as accessories. Analogous to cable impedance interfaces, optical time domain reflectometers may have interchangeable light sources, catering for different light wavelengths and fibre types. Chart recorders, too, are available; as are interfaces to specific networks.

10 Automatic test equipment

It used to be that the testing of electronic appliances was just a case of measuring a few independent analog measurand parameters, such as voltage and current amplitude, frequency and time relationships etc., at a small number of points. Generally the parameters could be measured one at a time, without problem.

Typical modern electronic appliances, on the other hand, are of a microprocessor-based system nature and demand testing of a large number of digital and analog measurand parameters at a correspondingly large number of points. Additionally, the parameters are often so interdependent that their values only have significance when monitored in relation with each other. Thus, measurements must be taken simultaneously.

Trends in electronic test equipment naturally reflect this change and Figure 10.1 illustrates the general move from single-time, single-measurand test instruments to multi-time, multi-measurand instruments. Simple analog and digital meters represent the basic equipment, capable of performing a single measurement at a single time. The oscilloscope extends the measurement by performing it over a period of time. Dual- and four-trace oscilloscopes allow a small number of measurements to be made over this period of time. Logic analysers take this facility two stages further: first, by allowing a large number of measurements to be made over the single period of time; and second, with recent developments in logic analysis, by allowing a large number of measurements to be made over a number of time

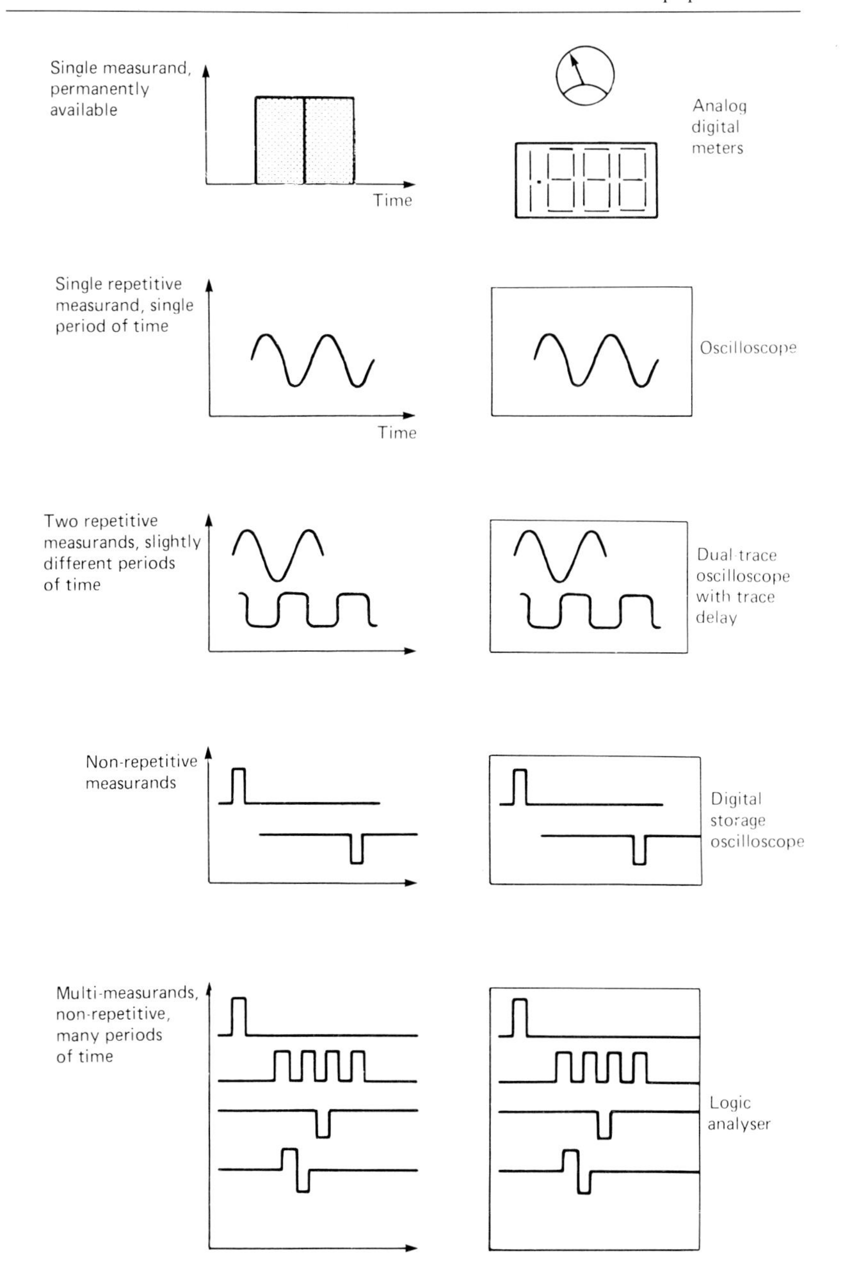

Figure 10.1 Trends in modern test equipment correspond to a general move from single-time, single-measurand instruments to multi-time, multi- measurand instruments

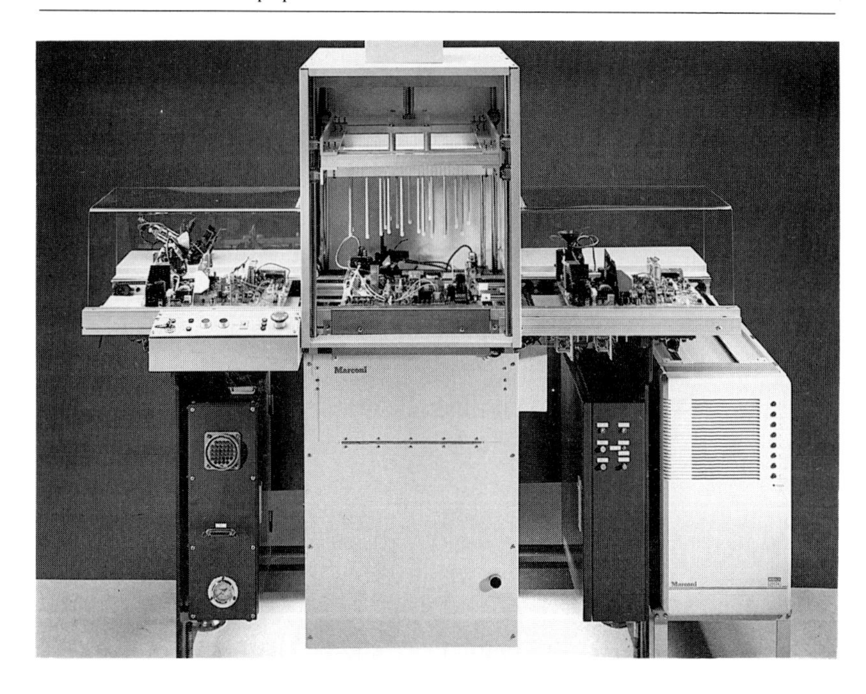

Photo 10.1 Midata 520 manufacturing defects analyser (Marconi Instruments)

periods. In all these instances, the user effectively controls the functions of the test instrument.

However, the trends do not stop there. As electronic appliances become even more complex and as manufacturers attempt to build in greater levels of reliability, so the test instruments used must themselves become even more complex. The limits suggested by Figure 10.1 are not, in fact, test equipment limitations but human limitations. It becomes increasingly difficult to correlate all the information regarding the many measurements which modern test equipment is capable of taking and, in many instances, is even impossible. For this reason much modern test equipment is microprocessor controlled.

An instrument, say a spectrum analyser, with microprocessor control includes many different functions which, opposed to a conventional spectrum analyser, are automatic. In the microprocessor-controlled spectrum analyser functions such as frequency scan, centre frequency, and resolution bandwidth may be automatically controlled: the microprocessor-controlled spectrum analyser is an example of an **automatic test instrument.** Most modern automatic test instruments are **programmable;** that is, they feature an interface which allows their internal microprocessor (and therefore their measurement functions) to be controlled by an external computer. Most, if not all, measurement facilities of an automatic test instrument may be set by a computer via this interface, and measurements taken are similarly relayed back to the computer for correlation and display. When a computer is used to control one or more programmable automatic test instruments the resultant system is what we know as **automatic test equipment** (ATE). This difference is important: automatic test instruments are devices capable of performing and displaying measurements autonomously or in a system; automatic test equipment is a complete measurement system, consisting of one or more automatic test instruments and a computer controller.

Such automatic test equipment requires computer control to ensure correct operation, record the measurements, and correlate the vast amounts of measurement data, presenting it in a form readily understood by the human user. In effect, the user no longer *directly* controls the test equipment (although the user must still program the computer which *does* control it), and most, if not all, functions are automatic.

Measurements are not limited by the user: any number of different measurements can be performed in any number of time periods. For example, the user of an analog voltmeter has great difficulty in taking and recording even one measurement a second.

Programmed automatic test equipment could take, record and display as many as one thousand measurements in the same time. Alternatively, automatic test equipment could take and record one measurement every second for the next thousand days – non-stop and accurately (without food, drink or sleep!).

Basic methods of automatic test equipment

There are two basic methods of creating automatic test equipment. First, a unique device may be designed and made, specifically for the purpose. Figure 10.2 shows an example of such a device, capable of taking a number of measurements of voltage and current, while counting events, measuring frequency, distortion and frequency response, and monitoring signals on a data bus. Output from the device to the system under test is a swept sine-wave signal. Control of the various measuring facilities is provided by the microprocessor-based heart of the

Figure 10.2 One method of creating automatic test equipment, in which the equipment is a unique device, built for one application

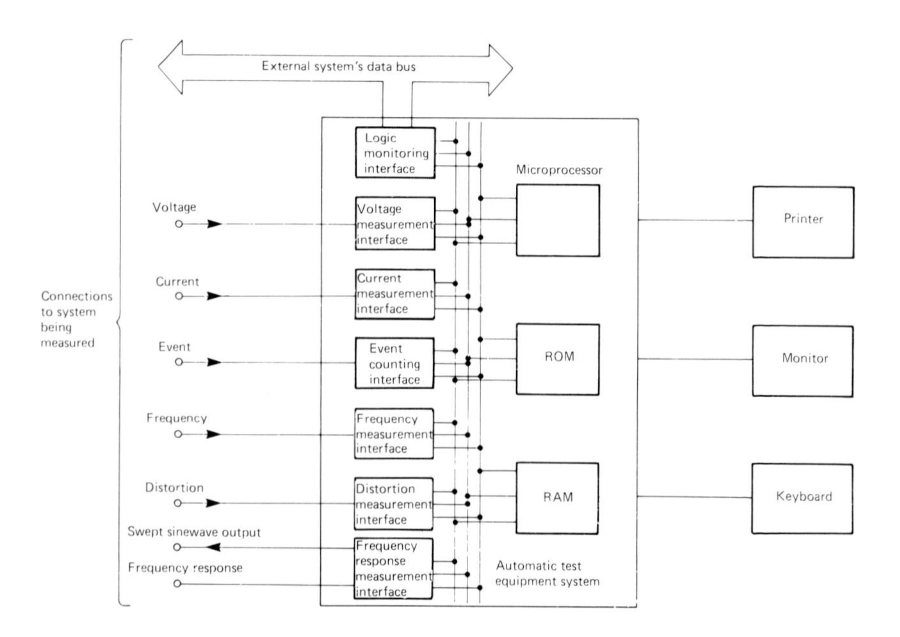

device, which in turn is controlled by programmed instructions from the user. This type of automatic test equipment is, in fact, a computer system complete with the necessary input and output units to allow measurements of the various parameters of the system under test. Recording of the values of these parameters and the format of the correlated information again depends on the user's programmed instructions, and is displayed either on a monitor or hard-copied onto paper with a printer.

The second method of creating automatic test equipment is to use a microprocessor-based personal computer to control general-purpose test equipment such as meters, universal counter timers, logic analysers, signal generators etc., as if they were peripheral devices, as illustrated in Figure 10.3. In this method each peripheral test equipment performs the measurements it is told to do by the computer on the system being monitored; it then relays the readings to the computer, which records and displays the correlated data, again on a monitor or printed onto paper. As in the first method, the user controls the overall system operation with programmed instructions.

From these descriptions it's fairly obvious what the main differences between the two methods are: the first is a custom-built *device*, which is likely to be quite expensive in terms of initial capital outlay, can only be used to test one particular system, and will most probably be used to test electronic appliances that are manufactured in vast quantities; the second is a custom-built *system*, which is still very expensive, but can be easily adapted to suit other test applications, so will most probably be used to test electronic appliances that are manufactured in quite small quantities (with the knowledge that the system can be easily adapted to suit other applications as and when necessary). In

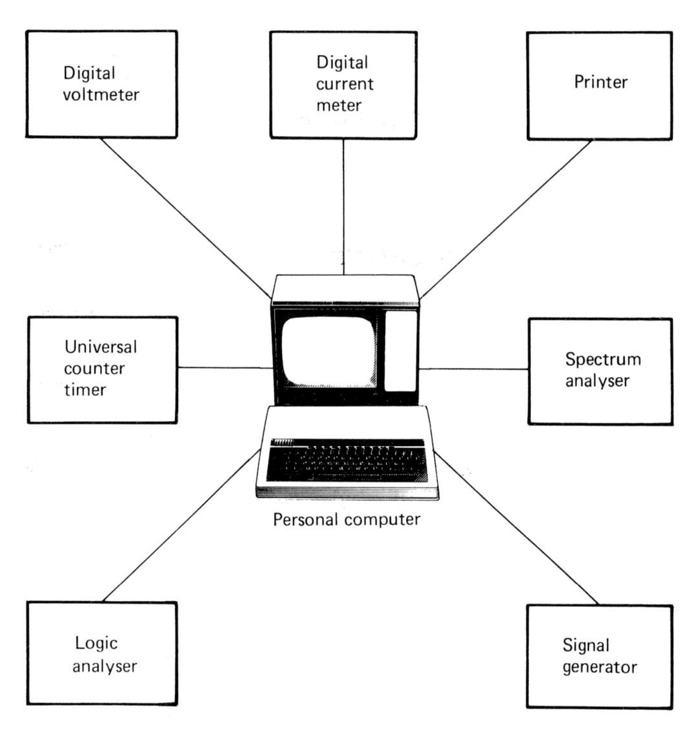

Figure 10.3 Another method of creating automatic test equipment, in which a personal computer is used to control programmable, readily available instruments

practice even the first method can usually be adapted, as a modular design approach is often used which allows the user to change measurement modules to suit other applications, and the two methods are simplified representations of extremes of automatic test equipment design philosophy. The philosophy, of course, is simply one of having a range of test instruments controlled by a computer which, in turn, is controlled by a user's program to suit the test application. In the first method above, the range of test instruments is built in to a complete computer-controlled device; in the second method the range is simply a collection of individual but interconnected instruments, controlled by a central computer.

If a number of instruments are to be controlled by a single, central computer it makes sense if each item's two-way interface links to a common data and control bus. Connections between the individual instruments and the computer are simplified enormously – often wiring is simply a matter of linking all the instruments with ribbon cable, as shown in Figure 10.4. A bus structure for automatic test equipment makes the system very flexible as extra instruments may be added with little fuss, and changing the system's test requirements is simply a matter of changing the instruments and reprogramming the computer. There are many bus standards in use and they are discussed in Chapter 11.

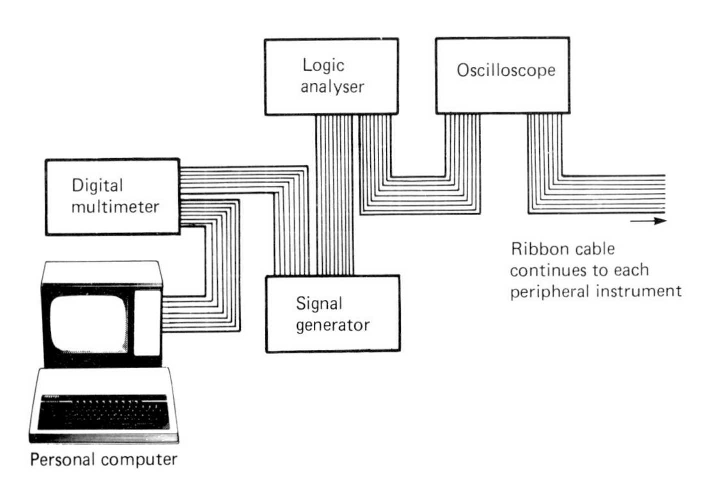

Figure 10.4 A common data and control bus, connected in parallel to each instrument using ribbon cable, makes a flexible and easily adapted automatic test equipment system

Recently a trend has developed, building on the automatic test equipment bus principle, in which the bus is controlled not by a stand-alone computer but by a purpose-built, standard-sized computer controller. Further, the peripheral equipment under computer control is also standard-sized and in modular form. A main benefit of this trend is that a single housing may then be used to hold all the modules. All functions and features of each peripheral equipment (including power supply) are controlled and adjusted by the computer via the bus.

Obviously system cost can be much lower (esti-

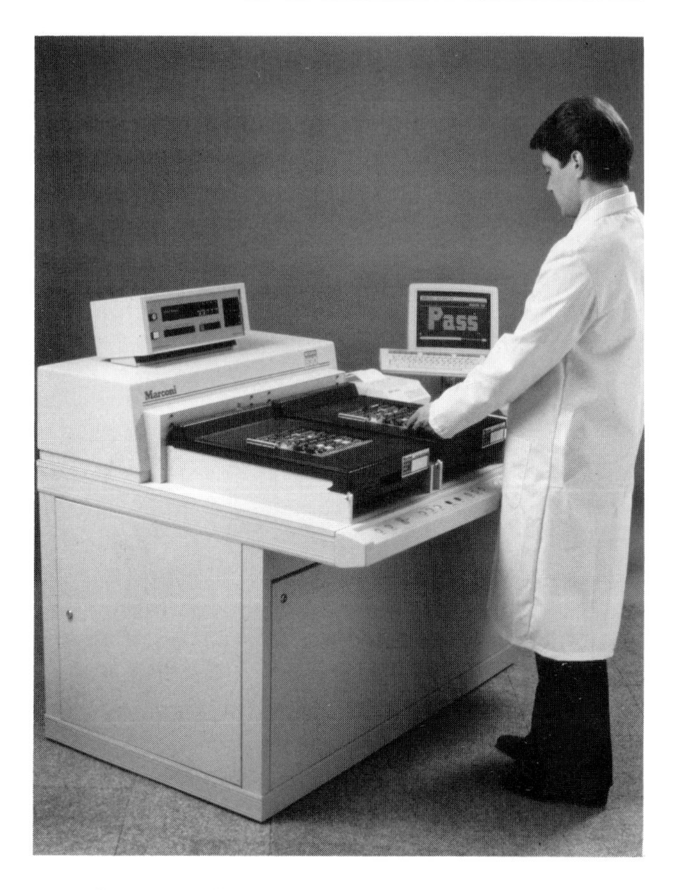

Photo 10.2 Midata 530 in-circuit tester (Marconi Instruments)

mated at around one third the price of an equivalent non-modular system), and overall size is considerably reduced (about one-tenth of the equivalent non-modular system). A modular system ensures that an automatic test equipment system is easily adaptable to future requirements: when a different system is needed, unwanted modules may be taken out of the housing and new modules simply slotted in.

Another important benefit is that a purpose-built bus may have extremely high data rates, a factor which is important where a large automatic test equipment system, with any measurands and programmed steps, is required.

We discuss this important trend towards modular test equipment in bus-structured form in Chapter 11.

Types of automatic test equipment

Complexity of automatic test equipment compared, say, with individual test instruments such as meters, oscilloscopes and logic analysers, is such that it is possible to itemise particular areas of involvement which necessitate specific equipment. Contrasted with individual test instruments, where an oscilloscope is an oscilloscope whether it's used on the bench or on the production line, automatic test equipment is often purpose-built for one application.

Generally, the applications where automatic test equipment is used to test a product parallel the

Photo 10.3 KTS2005 in-circuit tester (Measurement Limited)

normal stages of that product's life. Thus, automatic test equipment systems may be used in:

- design and development
- production
- reliability and certification test
- service

More often than not, different automatic test equipment systems are used at each stage, although it is possible to design a single system with the capability to test the product at every stage.

There is a number of types of automatic test equipment, broadly categorised into the main tests performed. There is, however, no reason why more than one type of test cannot be performed by a single system. Similarly, there is no reason why a single system cannot perform *all* tests required. Indeed, the trend is towards this. Use of a computer-controlled bus with modular peripheral instruments, as discussed here and in Chapter 11, makes such an encompassing automatic test equipment system possible.

Main categories of automatic test equipment include:

- **component testers** – to test individual parts prior to assembly within a product to ensure they fall within specified tolerances
- **unpackaged assembly testers** – to test assembled parts, say, printed circuits, prior to packaging
- **packaged assembly testers** – to test and ensure reliability of the complete and packaged product prior to use by a customer
- **maintenance and service equipment** – to repair and overhaul a used product.

These are very broad categories, though, and we must consider each in a little detail.

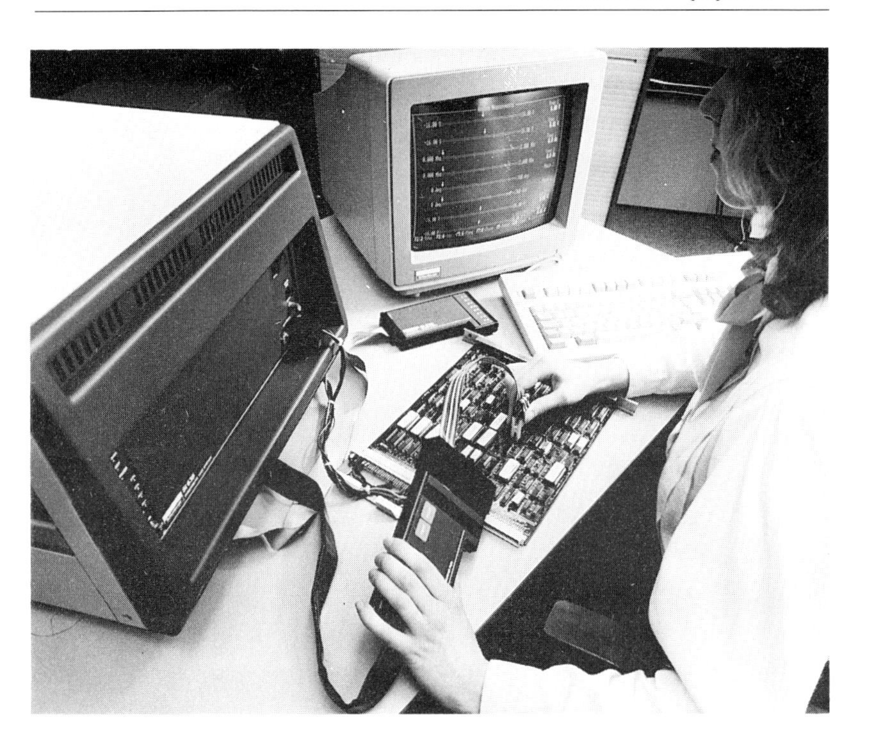

Photo 10.4 Solartron 635 functional tester (Schlumberger)

Component parts test equipment

Testing of component parts is rarely performed in anything other than manufacturing stages of a product. Nevertheless considerable test equipment is devoted to the testing of component parts and so must be considered.

Component parts testing is a fairly basic procedure. Generally, it is accomplished using simple procedures designed to determine measurands such as resistance, capacitance, semiconductor functions, dimensions, solderability and continuity. Testing is performed simply to ascertain the parts are of a specified quality. Tests inevitably depend on the parts.

Passive components are generally tested simply to measure their value and dimensions. Simple meter or bridge circuits can be used to perform such tests manually or semi-automatically. More complex, automatic measurements may be made using instruments which incorporate bridges, analog-to-digital conversion, function generators, voltage supplies, analog and digital stimuli and so on.

Active components are usually tested on a functional basis, that is, they are supplied with stimuli which simulate operating conditions, and resultant relationships are measured and compared with the ideal. Again, this may be a manual, semi-automatic or automatic test.

Printed circuit boards are often tested prior to asembly. Such **bare boards** can be of a range of complexity; from simple single-sided boards which will hold only a handful of through-hole (that is, leaded) components, to extremely complex multilayered boards (perhaps with over thirty internal track layers) designed to hold hundreds of surface mounted (that is, leadless) components. Consequently, test equipment used depends largely on the circuit boards to be tested.

Simple boards may require just a straightforward visual check, perhaps with a magnifying aid. Visual checks are not reliable, though, on more complex boards and so test equipment which checks continuity of board track is common. A simultaneous check of insulation between tracks is recommended, too. Automatic test equipment (placing the bare-board on a bed-of-nails fixture or, with a pair of moving probes programmed to position at large numbers of test points around the board) are available. With multilayered boards it is often impossible, though, to check internal layers for continuity.

Automatic optical inspection (AOI) using cameras, scanning lasers, or sometimes X-rays, are used to compare bare-board tracks with an ideal image (often called the **golden image**). Such systems, however sophisticated, cannot absolutely guarantee continuity.

Similar to the requirements of continuity in circuit boards, cable harnesses and backplanes of complex multi-board products need to be checked prior to assembly. Similar continuity checking test equipment is therefore available, too. It is becoming increasingly popular to use time domain reflectometers to provide a graphic signature of the harness or backplane, indicating presence and type of fault or similarity to a golden signature.

There is a method which sidesteps the requirement for testing component parts of a product – to use parts which are known and guaranteed to be of the required quality. National, regional and international standards organisations have coordinated (and continue to do so) standards and procedures which ensure components manufactured by a supplier are

Photo 10.5 Midata 590 functional tester (Marconi Instruments)

Photo 10.6 Rohde & Schwarz TSIC combinational test station (Rohde & Schwarz)

of a defined quality. Standards and procedures effectvely form complete specification systems, incorporating all types of components and manufacturers.

BS 9000 is the British specification system, CECC system operates in the UK and Europe, while ICEQ system operates in the UK, Europe and worldwide. Approved components are listed in frequently updated **qualified products lists** (QPLs) and it is a simple matter for the purchaser to identify the required components from these lists prior to purchasing.

This sort of self-assessment procedure, known commonly as **vendor assessment**, can aid product manufacture enormously. Accurate manufacturing times can be predicted and, overall, considerable wasted time may be eliminated. When component parts are purchased outside of such a procedure, **purchaser assessment** remains the only viable method of assuring a finished product's quality. Fortunately, test equipment to carry this out is easily available and not too expensive.

Unpackaged assembly test equipment

Once assembled and soldered, but prior to packaging, it is usual to test printed circuit boards to ensure the complete assembly performs as expected. Test equipment to do this could be used in maintenance and service environments, too, but is usually restricted to manufacturing stages simply because of large size.

Three main types of test may be undertaken: optical; in-circuit; functional. Optical inspection (particularly in an automated form) is possible using visible light, laser, infra-red or X-ray techniques discussed for bare board testing. Correct placement of components, short circuits, open circuits, faulty soldered joints and so on can all be isolated using optical (particularly automated) inspection.

In-circuit testing is performed by accessing nodal points within an assembly, generally with the use of a bed-of-nails fixture, then testing individual parts of the circuit, often to a component level. By comparing the measured values with defined ideal values, faults such as short circuits, misplaced components, wrongly valued components, poor soldered joints and defective tracks can be isolated. Tests on components are performed sequentially so that, depending on circuit complexity and numbers of components, a complete procedure may take considerable time. Nevertheless, it has been reported that some 90 per cent of manufacturing faults can be detected by in-circuit testing, so it can represent an extremely powerful procedure. However, it is not capable of testing overall performance.

Inherent in the use of in-circuit testing is the assumption that problems occur in the manufacturing stage – in other words they are manufacturing defects. For this reason certain in-circuit testing techniques are known as **manufacturing defects analysis**, although this term strictly refers to in-circuit testing when no power is applied to the assembly, testing the assembly passively. Manufacturing defects analysers are correspondingly cheaper than pure in-circuit testers, though cannot check individual active devices for correct function. Another name given to in-circuit testing is **pre-screening.**

Functional testing refers to a test procedure which involves the application of power and test signal patterns which simulate normal operation of the assembly, allowing a check of overall performance to be made. In this respect, functional testing is usually more rapid than in-circuit testing, and is often a simple pass or fail test (sometimes called **go/no go** testing) which can be of benefit in high volume production of assemblies. On the other hand, little indication of a specific fault is given compared with in-circuit testing. Access to test nodes of the assembly is through the assembly's board connectors.

Combination testers are available which perform both in-circuit and functional tests.

Packaged assembly test equipment

Once packaged in its housing a circuit board assembly may be subjected to tests to determine performance under stress conditions. These are generally of two types. First is a test procedure which simply determines whether the product operates under the stresses. Sometimes customer specifications (particularly of military origin) call for such stress tests and detail exact methods – these are known as **certification** or **qualification** tests.

Second, perhaps of greater potential, are tests designed to force failures to occur prior to delivery to the customer. Such failures, which would otherwise occur during the product's working life (during the early cycle of a product's reliability curve) are forced to occur during the tests by abnormal stresses imposed, including in order of effectiveness (Institute of Environmental Sciences):

- temperature cycling
- vibration
- high temperature
- electrical stress
- thermal shock
- sine vibration, fixed frequency
- low temperature
- sine vibration, sweep frequency
- combined environment
- mechanical shock
- humidity
- acceleration
- altitude

Tests such as these are known as **reliability screening tests**, or **environmental stress screening tests**.

Maintenance and service tests

Generally, maintenance and service tests are of a simpler nature than any of the tests detailed earlier. Service personnel can only take a limited amount of test equipment on-site, after all.

Tests which are used tend to be of a functional nature, simply to test products which have previously worked and for some reason have ceased to do so. Often these are returned as circuit board assemblies (replacing the faulty assembly with a known working assembly) to the manufacturing site, where it is known that complex in-circuit or functional test equipment, used to test assemblies on original manufacture, is located. Faults are in the main due to failure of single components, so in-circuit testing usually locates faults better than functional testing.

On-site, fairly basic test equipment prevents faults from being located rapidly, and time spent repairing faults under such conditions is not usually cost-effective, straightforward assembly replacement being a cheaper option.

Automatic test equipment administration
Complexity of much automatic test equipment is such that day-to-day running of the system represents a significant proportion (estimated at around 50 per cent – ERA Report 89/0353) of total investment. Put another way, whatever is spent purchasing automatic test equipment can be expected to be spent again maintaining, programming and adapting the system, and training users.

Maintenance of automatic test equipment
After an initial warranty period, it becomes the owner's responsibility to ensure equipment is adequately maintained. Maintenance falls into two main areas: service; and calibration.

Service is often aided by self-test routines incorporated into the test equipment, allowing users to check equipment performance. In such equipment the most common method is to connect a self-test adaptor to the equipment's fixture, then running the routines. Alternatively, test equipment manufacturers provide service support.

Calibration, where applicable, should be undertaken according to manufacturer's instructions. Sometimes calibration is automatic and no user involvement is required. Sometimes extra equipment is required to assist in calibration. Sometimes, equipment, or parts of it, must be returned to manufacturer for factory calibration.

Programming automatic test equipment
Whatever type or complexity of automatic test equip-

ment, a program is required to control it. Usually this is written in a high level computer language. Always the program is specific to the tested assembly-type – a new program is required to test a different assembly-type.

Photo 10.7 Midata 510 benchtop combinational tester (Marconi Instruments)

There are two basic methods of producing programs for automatic test equipment:

- manufacturer of the automatic test equipment provides a program service (or recommends an outside software house) to generate programs. This has the disadvantage that the equipment manufacturer does not have an in-depth knowledge of the assembly to be tested
- manufacturer of the assembly generates programs. Disadvantage here being the assembly manufacturer does not have an in-depth knowledge of the automatic test equipment. On the other hand, if the automatic test equipment is to be reprogrammed to test subsequent assemblies,

knowledge gained from the generation of the first program is available.

Adapting automatic test equipment

Fixtures of some description are used to adapt an automatic test equipment system to suit particular assemblies. For example, in-circuit testers, combinational testers and some bare board testers used bed-of-nails fixtures; while functional testers generally use printed circuit connectors.

A fixture is almost certainly assembly-specific, not usable for any other assembly-type. Consequently, design of a fixture is most likely to be done while the program is under generation, often by the programmer. A further consequence is that any minor changes in the assembly hardware must be matched by corresponding changes in the fixture.

Training users of automatic test equipment

Three levels of staff are likely to come in to contect with automatic test equipment: operators; maintenance engineers; and programmers.

Each level of staff requires training to be able to perform adequately. The more complex the test equipment, the more training is likely to be necessary for all levels.

Normally, automatic test equipment manufacturers provide training courses, usually on-site although, where test equiment is particularly complex, training may be given at the supplier's base.

Type and availability of automatic test equipment

A discussion of specific test equipment is beyond the scope of this book. Readers are referred to manufacturers' data, and to the report *Guide to low-cost ATE* (number: 89–0353) by ERA Technology Ltd, Cleeve Road, Leatherhead, Surrey KT22 7SA. The report details over fifty automatic test equipment systems costing under £50,000.

11
Buses

An automatic test equipment system is essentially a computer system. Peripheral instruments are computer-controlled, while data passes between peripherals and computer along one or more data buses.

Over recent years a number of types of data bus have been used to create automatic test equipment; all of them having different features which suit different applications. In some cases manufacturers designed a data bus unique to their own instruments, consequently restricting users' choice of peripheral items. Many of the data buses were borrowed from other electronic fields, such as the V24 (based on the earlier IEEE: RS232C) standard serial data communications bus, and the S100, 6800, and Z800 parallel microprocessor communications buses. This situation naturally led to a general non-conformity between test equipment which deterred potential users; equipment from one manufacturer could rarely be used alongside another manufacturer's equipment in the same automatic test equipment system.

In 1965, however, a data bus was defined by Hewlett Packard (the **Hewlett Packard interface bus** – HP-IB) specifically for the purpose of interfacing programmable measuring instruments, accessories and a computer into an automatic test equipment system. It was a more-or-less comprehensive bus which allowed all types of programmable measuring instruments to be connected into a single automatic system, so many other manufacturers incorporated it into their instruments during the early 1970s. For the first time, users could purchase instruments from many manufacturers and be certain they could be successfully interfaced.

The American Institute of Electrical and Electronic Engineers adopted the interface into its standard, the **IEEE 488**, ten years after it was first developed and American National Standards Institute incorporated it as the standard **ANSI MC 1.1**. Finally, the International Electrotechnical Commission adopted it in standard **IEC 625** (**BS 6146** in the UK), defining a different connector plug. All standards are fundamentally compatible so it became a worldwide instrument bus standard, referred to by any of these standard numbers or, popularly, a new name which avoids any ambiguity; the **general-purpose interface bus**, (GPIB).

Adoption of a GPIB standard effectively revolutionised automatic test equipment. Most automatic test instrument manufacturers produce instruments which have an interface to the GPIB, which means the instruments are programmable and may be used by themselves to measure and display readings of a system under test, or as part of an automatic test equipment system controlled by a computer.

GPIB does have its limitations, on the other hand. First, system data speed is a maximum of around 1 Mbyte s^{-1}, although in most applications much lower than this because overall speed depends on the slowest instrument connected to the bus. Second, a measurement system using the GPIB may comprise many peripheral instruments, all taking considerable workbench area.

Despite its limitations, though, such a standard at least made instrument manufacturers realise it is possible to interface peripheral instruments together to construct systems specifically suiting varied applications which users need. All that was needed was to take the concept one stage further, developing a bus which allowed modular instruments from any manufacturer to be added to an instrument bus within a single housing. In this way the two disadvantages of the GPIB: speed and size, are overcome.

This step was taken in 1987 by a consortium of test equipment manufacturers comprising Colorado Data Systems, Hewlett Packard, Racal-Dana, Tektronix and Wavetek. Realising the potential of such an instrument bus, the consortium proposed adoption of an existing computer bus (the VMEbus), upgrading and extending it to include specifications of module size and performance. The result – the **VXIbus** (short for *VMEbus extensions for instrumentation*) forms the basis of complete, high-speed, automatic test equipment systems in modular form. VXIbus data transfer rates are in the region of 1 Gbyte s^{-1}, and modules are in a range of standard sizes. Looking to avoid incompatibility with existing equipment where possible, the bus was designed to allow fairly simple interface modules to be used to connect GPIB or VMEbus instruments. Since then, over 100 manufacturers have registered with the consortium and so are licenced to build equipment to use with VXIbus.

For the next few years, at least, VXIbus looks set to become the basis of most automatic test equipment systems. Apart from advantages of size and speed over other bus systems, it has been quoted that an automatic test equipment system using VXIbus will be around one third the price of a similar system using one of the other buses and about one-tenth the size.

Bus data communications levels

Inside any computer system there are several levels at which bus communications can occur. Considering a range of systems, the main levels are:

- between integrated circuits on a printed circuit board
- between printed circuit boards in an instrument
- between instruments
- between a controller and peripheral instruments
- between separate systems, each with its own controller.

These levels are worth bearing in mind as we look at buses used in automatic test equipment systems. Understanding the levels of data communications taking place within bus systems will help us to understand the capabilities of each.

Available buses

There are many buses which have been used as the basis for automatic test equipment systems. Some of these are:

- V24/EIA232
- EIA422/EIA423
- EIA449
- S100
- Multibus
- PC expansion bus
- STE
- VMEbus
- GPIB
- VXIbus.

Table 11.1 lists these bus systems with the levels of data communications just described, together with numbers of address and data bits, maximum data rate, maximum range and system printed circuit board sizes. The first three buses are serial with data transmitted bit by bit. The remainder are parallel, with data transmitted byte by byte.

Table 11.1 Comparison of a range of buses used in automatic test equipment systems

Bus	*Level of data communications (between . . .)*	*Address bits*	*Data bits*	*Maximum speed (byte* s^{-1}*)*	*Range (m)*	*Circuit board size (mm)*
V24/EIA232	Controller and instruments	n/a	n/a	2K	20	n/a
EIA422/EIA423	Controller and instruments	n/a	n/a	100K	1000	n/a
EIA449	Controller and instruments	n/a	n/a	1M	1000	n/a
S100	Printed circuit boards controller and instruments	16/24	8/16	1M	10	114 by 165 Eurocard
Multibus	Printed circuit boards controller and instruments systems	24/32	8/16/32	10M	10	Eurocard
PC expansion bus	Printed circuit boards controller and instruments systems	20	8		n/a	
STE	Printed circuit boards instruments controller and instruments systems	20	8	5M	10	Eurocard
VMEbus	Printed circuit boards instruments controller and instruments systems	24/32	8/16/32	10M	10	Eurocard
GPIB	Instruments controller and instruments	32	8	1M	20	n/a
VIXbus	Printed circuit boards instruments controller and instruments systems	24/32	8/16/32	1G	10	Eurocard

V24/EIA232

The V24/EIA232 interface is used regularly to transfer data between computers, or between computers and peripheral instruments. Transfer is asynchronous or synchronous, and can be in a variety of operating modes: transmit only; receive only; half-duplex; full-duplex.

Users come into contact most regularly with the interface where computers or computer-based equipment transmit data over the analog telephone network using modems. Here the interface between the computer and modem is standardised worldwide by CCITT recommendation V24 (and in North America by ANSI EIA232 – which used to be called RS232 and often still is, incorrectly). In recommendation V24 the computer or equipment which generates and receives data is known as **data terminating equipment** (DTE), while the equipment which terminates the telephone line (the modem) is called **data circuit-terminating equipment** (DCE). Recommendation V24 defines the basic signal interchanges and functions between DTE and DCE; these are known as the **100 series** interchange circuits and are listed in Table 11.2. Where the modem automatically calls and answers, a further recommendation (V25) defines the extra circuits required which are known as the **200 series** interchange circuits and are listed in Table 11.3.

North American equivalent, EIA232, similarly defines the interface between DTE and DCE. Although the two have different designations, they are to all practical purposes interchangeable and equivalent. EIA232 interchange circuits are listed in Table 11.4. Figure 11.1 shows pin connections of the V24 25-pin D-connector.

EIA422/EIA423

Improving on the performance of the V24/EIA232

Table 11.2 V24 100 series interchange circuits

Interchange circuit		*Data*		*Control*		*Timing*	
Number	*Name*	*From DCE*	*To DCE*	*From DCE*	*To DCE*	*From DCE*	*To DCE*
101	Protective ground or earth						
102	Signal ground or common return						
103	Transmitted data		●				
104	Received data	●					
105	Request to send				●		
106	Ready for sending			●			
107	Data set ready			●			
108/1	Connect data set to line				●		
108/2	Data terminal ready				●		
109	Data channel received line signal detector			●			
110	Signal quality detector			●			
111	Data signalling rate selector (DTE)				●		
112	Data signalling rate selector (DCE)			●			
113	Transmitter signal element timing (DTE)						●
114	Transmitter signal element timing (DCE)					●	
115	Receiver signal element timing (DCE)					●	
116	Select standby				●		
117	Standby indicator			●			
118	Transmitted backward channel data		●				
119	Received backward channel data	●					
120	Transmit backward channel line signal				●		
121	Backward channel ready			●			
122	Backward channel received line signal detector			●			
123	Backward channel signal quality detector			●			
124	Select frequency groups				●		
125	Calling indicator			●			
126	Select transmit frequency				●		
127	Select receive frequency				●		
128	Receiver signal element timing (DTE)						●
129	Request to receive				●		
130	Transmit backward tone				●		
131	Received character timing					●	
132	Return to non-data mode				●		
133	Ready for receiving				●		
134	Received data present			●			
191	Transmitted voice answer				●		
192	Received voice answer			●			

interface, EIA422 and EIA423 provide better line matching which reduces reflections along the transmission line allowing higher data rates and line lengths to be used between DTE and DCE, while maintaining the same interchange circuits.

EIA422 specifies a balanced interface in which differential signal lines are used, terminated by an

Table 11.3 200 series interchange circuits

Interchange Circuit *Number*	*Name*	*From DCE*	*To DCE*
201	Signal ground	●	●
202	Call request		●
203	Data line occupied	●	
204	Distant station connected	●	
205	Abandon call	●	
206	Digit signal (2^0)		●
207	Digit signal (2^1)		●
208	Digit signal (2^2)		●
209	Digit signal (2^3)		●
210	Present next digit	●	
211	Digit present		●
213	Power indication	●	

Table 11.4 EIA232 interchange circuits

Interchange circuit		*Data*		*Control*		*Timing*	
Mnemonic	*Name*	*From DCE*	*To DCE*	*From DCE*	*To DCE*	*From DCE*	*To DCE*
AA	Protective ground						
AB	Signal ground/common return						
BA	Transmitted data		●				
BB	Received data	●					
CA	Request to send				●		
CB	Clear to send			●			
CC	Data set ready			●			
CD	Data terminal ready				●		
CE	Ring indicator			●			
CF	Received line signal detector			●			
CG	Signal quality detector			●			
CH	Data signal rate selector (DTE)				●		
CI	Data signal rate selector (DCE)			●			
DA	Transmitter signal element timing (DTE)						●
DB	Transmitter signal element timing (DCE)					●	
DD	Receiver signal element timing (DCE)					●	
SBA	Secondary transmitted data		●				
SBB	Secondary received data	●					
SCA	Secondary requet to send				●		
SCB	Secondary clear to send			●			
SCF	Secondary received line signal detector			●			

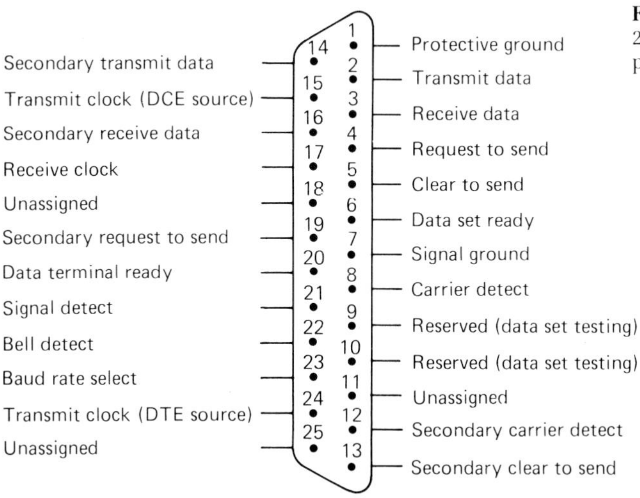

Figure 11.1 V24 25-pin connector pin assignments

impedance as low as 50 Ω. EIA423, on the other hand, specifies an unbalanced interface terminated by a 450 Ω impedance.

Specified matching terminations allow considerable performance improvements compared with the V24 interface. Data rates of around 100 Kbyte s^{-1}, and total distances of 1000 m are possible. Further, the interfaces allow more than one remote peripheral to be connected to the bus.

EIA449

Improving further on V24/EIA232 interface performance, EIA449 uses a different series of interchange circuits, listed in Table 11.5. Two connectors (a 37-pin for the primary channel and a 9-pin D-connector for the secondary channel) are used, and the interface is capable of data rates up to around 1 Mbyte s^{-1}.

S100

One of the first standardised computer buses, S100 was originally developed for 8080-based micro-

Table 11.5 EIA449 interchange circuits

	Interchange circuit		*Data*		*Control*		*Timing*	
	Mnemonic	*Name*	*From DCE*	*To DCE*	*From DCE*	*To DCE*	*From DCE*	*To DCE*
	SG	Signal ground						
	SC	Send common						
	RC	Receive common				●		
	IS	Terminal in service			●			
	IC	Incoming call			●			
	TR	Terminal ready				●		
	DM	Data mode			●			
Primary channel	SD	Send data		●				
Primary channel	RD	Receive data	●					
Primary channel	TT	Terminal timing						●
Primary channel	ST	Send timing					●	
Primary channel	RT	Receive timing					●	
Primary channel	RS	Request to send				●		
Primary channel	CS	Clear to send			●			
Primary channel	RR	Receiver ready			●			
Primary channel	SQ	Signal quality			●			
Primary channel	NS	New signal				●		
Primary channel	SF	Select frequency				●		
Primary channel	SR	Signalling rate selector				●		
Primary channel	SI	Signalling rate indicator			●			
Secondary channel	SSD	Secondary send data		●				
Secondary channel	SRD	Scondary receive data	●					
Secondary channel	SRS	Secondary request to send				●		
Secondary channel	SCS	Secondary clear to send			●			
Secondary channel	SRR	Secondary receiver ready			●			
	LL	Local loopback				●		
	RL	Remote loopback				●		
	TM	Test mode			●			
	SS	Select standby				●		
	SB	Standby indicator			●			

computer systems. It uses a card frame chassis with a backplane connected to double-sided 100-pin printed circuit board edge connectors, and boards are simply slotted into the chassis as required, creating a very adaptable system. Up to twenty-two boards may be used in a single system. The format has been used in many microprocessor-based computer systems since.

As the name indicates, S100 systems have 100 bus lines, all designated with particular purposes within a computer system, although not all need to be used in a given system. Originally all lines were unidirectional, input and output lines being separate, but a revision ganged the two 8-bit data in and out lines into a single 16-bit bidirectional bus. This and other inclusions are adopted in the EIA696 standard.

Overall structure of the bus system comprises:

- 16 data lines
- 24 address lines
- 19 control lines
- 8 interrupt lines
- 8 status lines
- 20 utility lines
- 5 power lines.

Table 11.6 lists pin assignments of the S100, while Figure 11.2 illustrates pin numbering used for the printed circuit board connector.

Table 11.6 S100 edge connector pin assignments

Pin no.	*Abbreviation*	*Signal/function*
1	+8 V	Unregulated supply rail
2	+18 V	Unregulated supply rail
3	XRDY	Ready input to bus master
4	VI0	Vectored interrupt line 0
5	VI1	Vectored interrupt line 1
6	VI2	Vectored interrupt line 2
7	VI3	Vectored interrupt line 3
8	VI4	Vectored interrupt line 4
9	VI5	Vectored interrupt line 5
10	VI6	Vectored interrupt line 6
11	VI7	Vectored interrupt line 7
12	NMI	Non-maskable interrupt
13	PWRFAIL	Power fail signal (pulled low when a power failure is detected)
14	DMA3	DMA request line with highest priority
15	A18	Extended address bus line 18
16	A16	Extended address bus line 16
17	A17	Extended address bus line 17

Pin no.	*Abbreviation*	*Signal/function*
18	SDSB	Status disable (tri-states all status lines)
19	CDSB	Command disable (tri-states all control input lines)
20	GND	Common 0 V line
21	NDEF	Undefined
22	ADSB	Address disable (tri-states all address lines)
23	DODSB	Data out disable (tri-states all data output lines)
24	0	Bus clock
25	ρSTVAL	Status valid strobe (indicates that status information is true)
26	ρHLDA	Hold acknowledge (signal from the current bus master which indicates that control will pass to the device seeking bus control on the next bus cycle)
27	RFU	Reserved for future use
28	RFU	Reserved for future use
29	A5	Address line 5
30	A4	Address line 4
31	A3	Address line 3
32	A15	Address line 15
33	A12	Address line 12
34	A9	Address line 9
35	DO1/Data 1	Data out line 1/bidirectional data line 1
36	DO0/Data 0	Data out line 0/bidirectional data line 0
37	A10	Address line 10
38	DO4/Data 4	Data out line 4/bidirectional data line 4
39	DO5/Data 5	Data out line 5/bidirectional data line 5
40	DO6/Data 6	Data out line 6/bidirectional data line 6
41	DI2/Data 10	Data in line 2/bidirectional data line 10
42	DI3/Data 11	Data in line 3/bidirectional data line 11

Pin no.	*Abbreviation*	*Signal/function*
43	DI7/Data 15	Data in line 7/bidirectional data line 15
44	sM1	M1 cycle (indicates that the current machine cycle is in an operation code fetch)
45	sOUT	Output (indicates that data is being transferred to an output device)
46	sINP	Input (indicates that data is being fetched from an intput device)
47	sMEMR	Memory read (indicates that the bus master is fetching data from memory)
48	sHLTA	Halt acknowledge (indicates that the bus master is executing an HLT instruction)
49	CLOCK	2 MHz clock
50	GND	Common 0 V
51	+8 V	Unregulated supply rail
52	−16 V	Unregulated supply rail
53	GND	Common 0 V
54	SLAVE CLR	Slave clear (resets all bus slaves)
55	DMA0	DMA request line (lowest priority)
56	DMA1	DMA request line
57	DMA2	DMA request line
58	sXTRQ	16-bit data request (requests slaves to assert SIXTN)
59	A19	Address line 19
60	SIXTN	16-bit data acknowledge (slave response to sXTRQ)
61	A20	Extended address bus line 20
62	A21	Extended address bus line 21
63	A22	Extended address bus line 22
64	A23	Extended address bus line 23
65	NDEF	Not defined
66	NDEF	Not defined
67	PHANTOM	Phantom (disables normal slaves and enables phantom slaves which share addresses with the normal set)
68	MWRT	Memory write
69	RFU	Reserved for future use

Pin no.	*Abbreviation*	*Signal/function*
70	GND	Common 0 V
71	RFU	Reserved for future use
72	RDY	Ready input to bus master
73	INT	Interrupt request
74	HOLD	Hold request (request from device wishing to have control of the bus)
75	RESET	Reset (resets bus master devices)
76	ρSYNC	Synchronising signal which indicates the first bus state of a bus cycle
77	ρWR	Write (indicates that the bus master has placed valid data on the DO bus /data bus)
78	ρDBIN	Data bus in (indicates that the bus master is requesting data on the DI bus/data bus)
79	A0	Address line 0
80	A1	Address line 1
81	A2	Address line 2
82	A6	Address line 6
83	A7	Address line 7
84	A8	Address line 8
85	A13	Address line 13
86	A14	Address line 14
87	A11	Address line 11
88	DO2/DATA 2	Data out line 2/bidirectional data line 2
89	DO3/DATA 3	Data out line 3/bidirectional data line 3
90	DO7/DATA 7	Data out line 7/bidirectional data line 7
91	DI4/DATA 12	Data in line 4/bidirectional data line 12
92	DI5/DATA 13	Data in line 5/bidirectional data line 13
93	DI6/DATA 14	Data in line 6/bidirectional data line 14
94	DI1/DATA 9	Data in line 1/bidirectional data line 9
95	DI0/DATRA 8	Data in line 0/bidirectional data line 8

Pin no.	*Abbreviation*	*Signal/function*
96	sINTA	Interrupt acknowledge
97	sWO	Write output (used to gate data from the bus master to a slave)
98	ERROR	Error (indicates that an error has occurred during the current bus cycle)
99	POC	Power on clear (clears all devices attached to the bus when power is first applied)
100	GND	Common 0 V

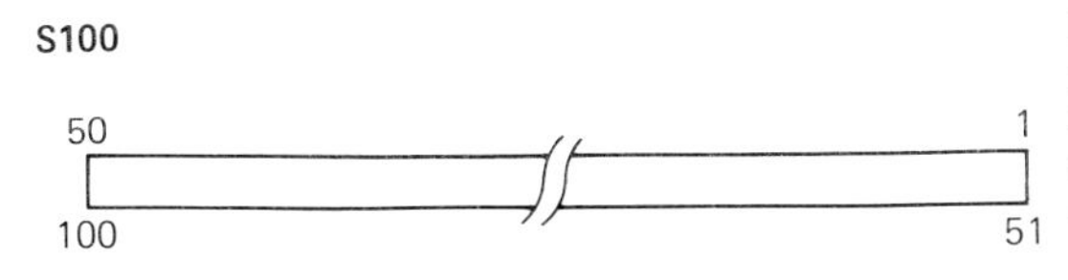

Figure 11.2 Pin numbering in the S100 printed circuit board connector

Multibus

Similar to the S100 bus, designed by Intel for its 8086 range of microprocessors, the multibus is an 86-line bus based on printed circuit board edge connectors. Overall bus structure comprises:

- 16 data lines
- 20 address lines
- 8 interrupt lines
- 17 control lines
- 5 power lines
- 2 lines reserved for future use.

Some power lines are found on more than one pin of the edge connector.

Table 11.7 lists pin assignments of the multibus, while Figure 11.3 illustrates pin numbering of the edge connector.

Table 11.7 Multibus edge connector pin assignents

Component side			
Pin no.	*Signal group*	*Abbreviation*	*Signal/function*
1	Supply rails	GND	Ground/common 0 V
3		+5 V	+5 V DC supply rail
5		+5 V	+5 V DC supply rail
7		+12 V	+12 V DC supply rail
9		−5 V	−5 V DC supply rail
11		GND	Ground/common 0 V
13	Bus control	BCLK	Bus clock
15		BPRN	Bus priority input
17		BUSY	Bus busy
19		MRDC	Memory read command
21		IORC	I/O read command
23		XACK	Transfer acknowledge
25			Reserved
27		BHEN	Byte high enable
29		CBRQ	Common bus request
31		CCLK	Constant clock
33		INTA	Interrupt acknowledge
35	Interrupt	INT6	Parallel interrupt requests
37		INT4	
39		INT2	
41		INT0	
43	Address bus	ADRE	Address line
45		ADRC	
47		ADRA	
49		ADR8	
51		ADR6	
53		ADR4	
55		ADR2	
57		ADR0	
59	Data bus	DATE	Data lines
61		DATC	
63		DATA	
65		DAT8	
67		DAT6	
69		DAT4	

Component side

Pin no.	*Signal group*	*Abbreviation*	*Signal/function*
71		DAT2	
73		DAT0	
75	Supply	GND	Ground/Common 0 V
77	rails		Reserved
79		−12 V	−12 V DC supply rail
81		+5 V	+5 V DC supply rail
83		+5 V	+5 V DC supply rail
85		GND	Ground/common 0 V

Track side

Pin no.	*Signal group*	*Abbreviation*	*Signal/function*
2	Supply	GND	Ground/common 0 V
3	rails	+5 V	+5 V DC supply rail
6		+5 V	+5 V DC supply rail
8		+12 V	+12 V DC supply rail
10		−5 V	−5 V DC supply rail
12		GND	Ground/common 0 V
14	Bus	INIT	Initialise
16	control	BPRO	Bus priority output
18		BREQ	Bus request
20		MWTC	Memory write command
22		IOWC	I/O write command
24		INH1	Inhibit 1 (disable RAM)
26		INH2	Inhibit 2 (disable ROM)
28	Address	AD10	Address lines
30	bus	AD11	
32		AD12	
34		AD13	
36	Interrupt	INT7	Parallel interrupt requests
38		INT5	
40		INT3	
42		INT1	
44	Address	ADRF	Address lines
46	Bus	ADRD	
48		ADRB	
50		ADR9	

Track side *Pin no.*	*Signal group*	*Abbreviation*	*Signal/function*
52		ADR7	
54		ADR5	
56		ADR3	
58		ADR1	
60	Data	DATF	Data lines
62	bus	DATD	
64		DATB	
66		DAT9	
68		DAT7	
70		DAT5	
72		DAT3	
74		DAT1	
76	Supply	GND	Ground/common 0 V
78	rails		Reserved
80		−12 V	−12 V DC supply rail
82		+5 V	+5 V DC supply rail
84		+5 V	+5 V DC supply rail
86		GND	Ground-common 0 V

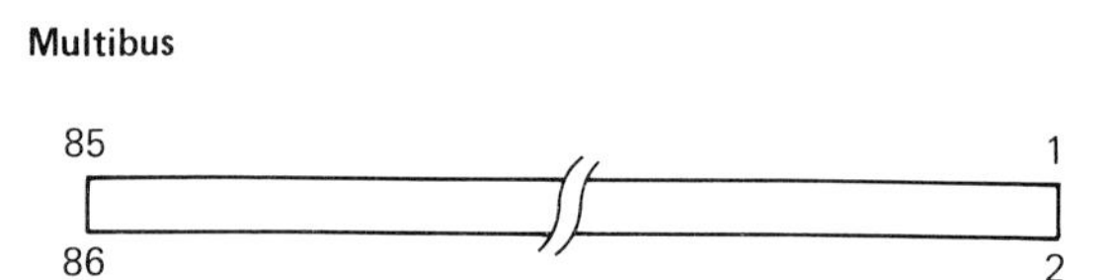

Figure 11.3 Pin numbering in the Multibus printed circuit board connector

PC expansion bus

IBM PC is an extremely popular computer, and it is no surprise that it has been used as the central control of automatic test equipment systems. Generally, this is accomplished using the PC expansion bus, which features a 62-pin printed circuit board edge connector.

Overall bus structure comprises:

- 8 data lines
- 20 address lines

- 8 interrupt lines
- 18 control lines
- 5 power lines
- 1 line reserved for future use.

Table 11.8 lists pin assignments of the PC expansion bus, while Figure 11.4 illustrates pin numbering of the edge connector.

Table 11.8 IBM PC expansion bus edge connector pin assignments

Pin no.	*Abbreviation*	*Signal/function*
1	GND	Ground/common 0 V
2	CHCK	Channel check output (when low this indicates that some form of error has occurred)
3	RESET	Reset (when high this line resets all expansion cards)
4	D7	Data line 7
5	+5 V	+5 V DC supply rail
6	D6	Data line 6
7	IRQ2	Interrupt request input 2
8	D5	Data line 5
9	−5 V	−5 V DC supply rail
10	D4	Data line 4
11	DRQ2	DMA request input 2
12	D3	Data line 3
13	−12 V	−12 V DC supply rail
14	D2	Data line 2
15		Reserved
16	D1	Data line 1
17	+12 V	+12 V DC supply rail
18	D0	Data line 0
19	GND	Ground/common 0 V
20	BCRDY	Ready input (normally high, pulled low by a slow memory or I/O device to signal that it is not ready for data transfer to take place)
21	IMW	Memory write output
22	AEN	Address enable output
23	IMR	Memory read output

Pin no.	*Abbreviation*	*Signal/function*
24	A19	Address line 19
25	IIOW	I/O write output
26	A18	Address line 18
27	IIOR	I/O read output
28	A17	Address line 17
29	DACK3	DMA acknowledge output 3 (see notes)
30	A16	Address line 16
31	DRQ3	DMA request input 3
32	A15	Address line 15
33	DACK1	DMA acknowledge output 1 (see notes)
34	A14	Address line 14
35	DRQ1	DMA request input 1
36	A13	Address line 13
37	DACK0	DMA acknowledge output 0 (see notes)
38	A12	Address line 12
39	XCLK4	4 MHz clock (CPU clock divided by two 200 ns period, 50% duty cycle)
40	A11	Address line 11
41	IRQ7	Interrupt request line 7 (see notes)
42	A10	Address line 10
43	IRQ6	Interrupt request line 6 (see notes)
44	A9	Address line 9
45	IRQ5	Interrupt request line 5
46	A8	Address line 8
47	IRQ4	Interrupt request line 4 (see notes)
48	A7	Address line 7
49	IRQ3	Interrupt request line 3
50	A6	Address line 6
51	DACK2	DMA acknowledge 2
52	A5	Address line 5
53	TC	Terminal count output (pulsed high to indicate that the terminal count for a DMA transfer has been reached)

Pin no.	*Abbreviation*	*Signal/function*
54	A4	Address line 4
55	ALE	Address latch enable output
56	A3	Address line 3
57	+5 V	+5 V DC supply rail
58	A2	Address line 2
59	14 MHz	14.31818 MHz clock (fast clock with 70 ns period, 50% duty cycle)
60	A1	Address line 1
61	GND	Ground/common OV
62	IA0	Address line 0

Notes: (a) Signal direction is quoted relative to the motherboard
(b) IRQ4 is generated by the motherboard serial interface
IRQ6 is generated by the motherboard disk interface
IRQ7 is generated by the motherboard parallel interface
(c) DACK0 is used to refresh dynamic memory, while DACK1 to DACK3 are used to acknowledge DMA requests.

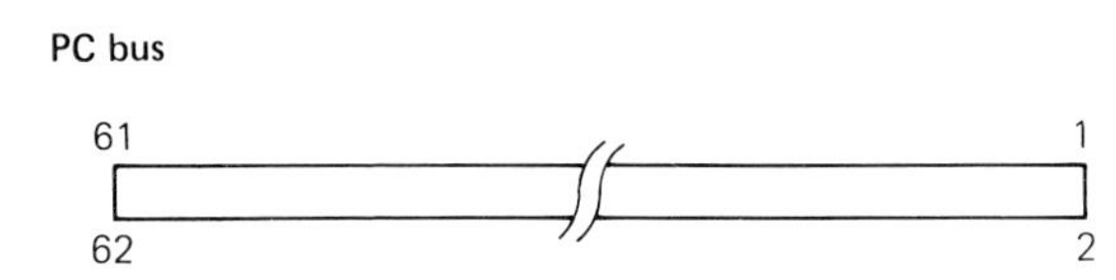

Figure 11.4 Pin numbering of the IBM PC expansion bus printed circuit board connector

STE

STE is the successor to an earlier bus; STD, upgraded and redesigned to allow modular backplaned construction using standard Eurocard-sized modules.

Original STD bus was produced in 1978 and is specifically for 8-bit microprocessors, and so is fairly limited. Nevertheless, STE bus features a 20-bit address bus and a maximum data rate of around 5 Mbyte s^{-1}. It is more of an industrial control system than one for instrumentation, but can be and is used for automatic test equipment systems.

Overall bus structure includes:

- 8 data lines
- 20 address lines
- 20 control lines
- 4 power supply lines
- 1 clock line.

Connectors used for the modular construction are standard Eurocard connectors with three rows of pins, 32 pins in each row. This is the first bus system considered here which uses such connectors. Connector is shown in Figure 11.5, with pin assignments. Note the middle row (row B) of the connector is not used.

VMEbus

One of the most important buses designed specifically for modular computer systems is the VMEbus. Like many of the bus systems already discussed, it is a backplaned system, such that modular boards may be plugged into chassis mounted edge connectors. Pin connections to the edge connectors are such that a board may be plugged into the bus at any location within the chassis (slot 0, however, is reserved for the system controller board).

Although electrically and logically similar to the 68000 microprocessor-based computer system, VMEbus is an open system, not dependent on any particular microprocessor. Systems based on other microprocessors are, indeed, common.

Board size is standardised into two sizes; connectors found on a board depend primarily on size. Two sizes (A and B) are single- and double-height Eurocards. An A-sized board has a single connector, known as the P1 connector. B-sized boards must

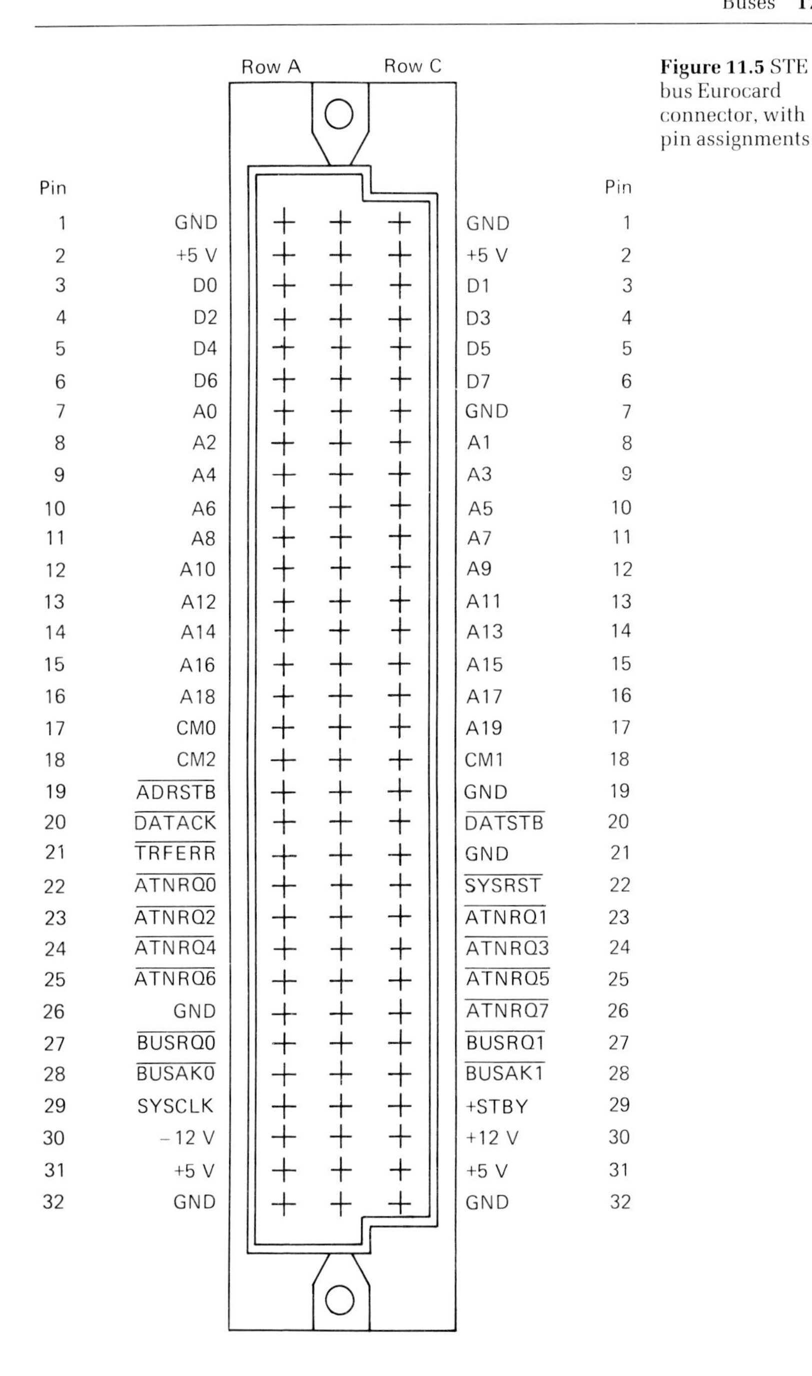

Figure 11.5 STE bus Eurocard connector, with pin assignments

have the P1 connector, with an optional P2 connector. Slot 0 in the chassis is assigned to a system controller, while remainder of the twenty-one possible slots of the bus system may be taken by any combination of A- and B-sized boards of whatever function.

Connectors are the standard Eurocard connectors, comprising three rows of 32-pins each on a 1/10 inch spacing, as used in the STE bus. Connector P1 pins are all assigned and are listed in Table 11.9. Table 11.10 lists pin assignments for the optional P2 connector – only the central row is defined, other two rows are user-definable. Often, these undefined pins are used for interface connections, say, to access an internal disk drive.

Overall bus structure for a system using only P1 connectors is:

- 16 data lines
- 24 address lines
- 7 interrupt lines
- 42 control lines
- 14 power lines.

Bus structure is extended if P2 connectors are used, by a further:

- 16 data lines
- 8 address lines
- 1 line reserved for future use.

GPIB

GPIB standards define all hardware characteristics (that is, cable, connectors, voltage and current values of signals, purposes of the signals, and timing relationships between the signals) of the interface, but leave software characteristics to the user. In

Table 11.9 VMEbus P1 connector pin assignments

Pin number	*Row a signal mnemonic*	*Row b signal mnemonic*	*Row c signal mnemonic*
1	DO0	BBSY*	D08
2	DO1	BCLR*	D09
3	DO2	ACFAIL*	D10
4	DO3	**BGOIN***	D11
5	DO4	**BG0OUT***	D12
6	DO5	**BG1IN***	D13
7	DO6	**BG1OUT***	D14
8	DO7	**BG2IN***	D15
9	**GND**	**BG2OUT***	**GND**
10	SYSCLK	**BG3IN***	SYSFAIL*
11	**GND**	**BG3OUT***	BERR*
12	DSI*	BR0*	SYSRESET*
13	DSO*	BR1*	LWORD*
14	WRITE*	BR2*	AM5
15	**GND**	BR3*	A23
16	DTACK*	AMO	A22
17	**GND**	AM1	A21
18	AS*	AM2	A20
19	**GND**	AM3	A19
20	IACK*	**GND**	A18
21	**IACKIN***	SERCLK(1)	A17
22	**IACKOUT***	SERDAT*(1)	A16
23	AM4	**GND**	A15
24	AO7	IRQ7*	A14
25	AO6	IRQ6*	A13
26	AO5	IRQ5*	A12
27	AO4	IRQ4*	A11
28	AO3	IRQ3*	A10
29	AO2	IRQ2*	AO9
30	AO1	IRQ1*	AO8
31	**−12 V**	**+ 5VSTDBY**	**+12 V**
32	**+5 V**	**+5 V**	**+5 V**

* Designates an active low signal

Note: Signal assignments shown in bold print in above table **are not** terminated. All remaining 72 signal lines **are** terminated.

Table 11.10 VMEbus optional P2 connector pin assignments

Pin number	*Row a signal mnemonic*	*Row b signal mnemonic*	*Row c signal mnemonic*
1	User defined	**+5 V**	User defined
2	User defined	**GND**	User defined
3	User defined	RESERVED	User defined
4	User defined	A24	User defined
5	User defined	A25	User defined
6	User defined	A26	User defined
7	User defined	A27	User defined
8	User defined	A28	User defined
9	User defined	A29	User defined
10	User defined	A30	User defined
11	User defined	A31	User defined
12	User defined	**GND**	User defined
13	User defined	**+5 V**	User defined
14	User defined	D16	User defined
15	User defined	D17	User defined
16	User defined	D18	User defined
17	User defined	D19	User defined
18	User defined	D20	User defined
19	User defined	D21	User defined
20	User defined	D22	User defined
21	User defined	D23	User defined
22	User defined	**GND**	User defined
23	User defined	D24	User defined
24	User defined	D25	User defined
25	User defined	D26	User defined
26	User defined	D27	User defined
27	User defined	D28	User defined
28	User defined	D29	User defined
29	User defined	D30	User defined
30	User defined	D31	User defined
31	User defined	**GND**	User defined
32	User defined	**+5 V**	User defined

Note: Signal assignments shown in bold print in table above are **not** terminated. All remaining 25 signal lines **are** terminated.

other words, connecting the automatic test equipment system components together is simply a matter of plugging them in, but controlling the system operation relies on the user buying or writing the software required to perform the measurement task.

Figure 11.6 illustrates the GPIB bus structure. All system components are connected in parallel to the cable of the bus, and all components have access to all lines making up the bus. Any component on the bus may (depending on its capabilities) send or receive data to or from any other component; also, a sender of data may send data, simultaneously, to more than one receiver. The components of a GIPB system fall into one of the following categories.

- a **listener** – can only receive data
- a **talker** – can only send data
- a **listener/talker** – can be switched between listening and talking
- a **controller** – the computer that determines which instruments talk and which listen during data transfers. The controller also has the task of sending special commands called **interface messages** to instruments on the bus. Only one controller is necessary on a bus.

The GPIB consists of 16 signal lines, divided into three groups comprising:

- 8 data lines for transfer of measurement data and addresses
- 3 lines for data transfer control
- 5 lines for interface management and control.

The eight data lines, DIO 1 to DIO 8, are bidirectional. Data is transferred byte by byte (that is, byte-serial), the eight bits of each byte being transferred in parallel (bit-parallel). Bytes are exchanged between

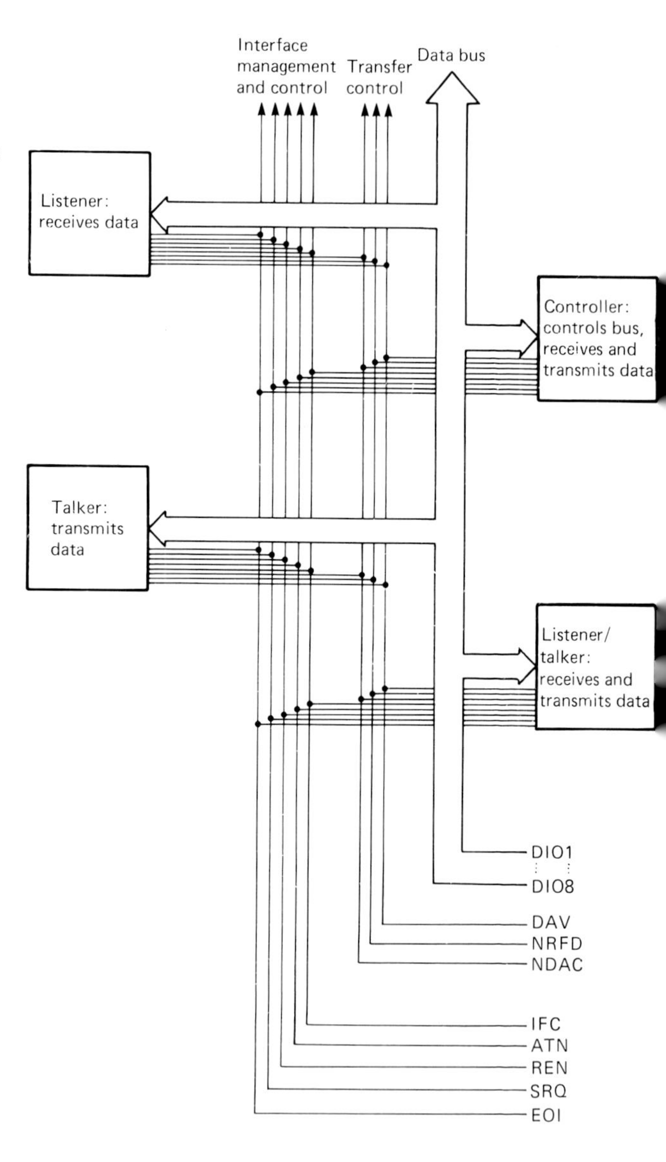

Figure 11.6 The GPIB bus structure. One controller manages all communications between instruments connected to the bus

an enabled talker and an enabled listener in a handshaking sequence, described later.

The three data transfer control lines are now described.

Data valid (DAV)
This signal indicates that valid data from a talker is present on the data lines, ready to be sent to listeners.

Not ready for data (NRFD)
This signal indicates that not all listeners are ready to receive data. Data transfer is only allowed to take place if all listeners indicate readiness to accept data; thus if any listener asserts the NRFD signal the talker is inhibited from asserting a DAV signal.

Not data accepted (NDAC)
This signal indicates that not all listeners have received the data. The signal only becomes unasserted when all listeners have accepted the data byte currently on the data lines.

The five interface management and control lines are now described.

Interface clear (IFC)
This signal can be asserted by the controller alone. The signal is used to set the interfaces of all peripheral units to a predefined condition. This is useful, say, immediately after switch-on.

Attention (ATN)
This signal indicate the nature of the information on the data lines. A logic 0 signal indicates the presence of data: a logic 1 indicates the presence of addresses or commands.

Remote enable (REN)
This signal instructs all instruments connected to the bus to be prepared for remote control operation. All panel controls of the instruments are blocked as soon as they are addressed as a listener.

Service request (SRQ)
A signal on this line allows any instrument to

interrupt the controller, thereby demanding attention.

End of identify (EOI)
Alone, this signal indicates the end of a multiple-byte transfer sequence. When EOI is sent in conjunction with ATN, the signals force the controller to execute a polling sequence which identifies the instrument requesting service.

Handshaking
Transferring data between one instrument and another is a reasonably simple task. The controller first assigns one instrument as a talker and one or more instruments as listeners, issuing talker and listener addresses over the bus. To do this the controller *attention* (ATN) signal line is first set to logic 1, thus informing all instruments that the data bus contains address information. After the talker and listener addresses have been issued the controller changes the ATN signal line to logic 0, and data transfer control is passed to the talker. The talker now simply places the data onto the bus and the listeners simply accept data in an asynchronous handshaking procedure, one byte at a time.

Figure 11.7 shows a timing diagram that illustrates the basic handshake procedure between talker and listeners on the GPIB (that is, after the controller has assigned one instrument as the talker, and one or more instruments as listeners). The bus operates with TTL signal levels in negative logic.

The controller first sends a *remote enable* (REN at t_1) and signals *attention* (ATN at t_2), which causes the instruments to set their *not ready for data* (NRFD) and *not data accepted* (NDAC) lines. At t_3 the fastest instrument is ready to accept data. However, as the *not ready for data* (NRFD) signals of all connected instruments are in AND logic, the line

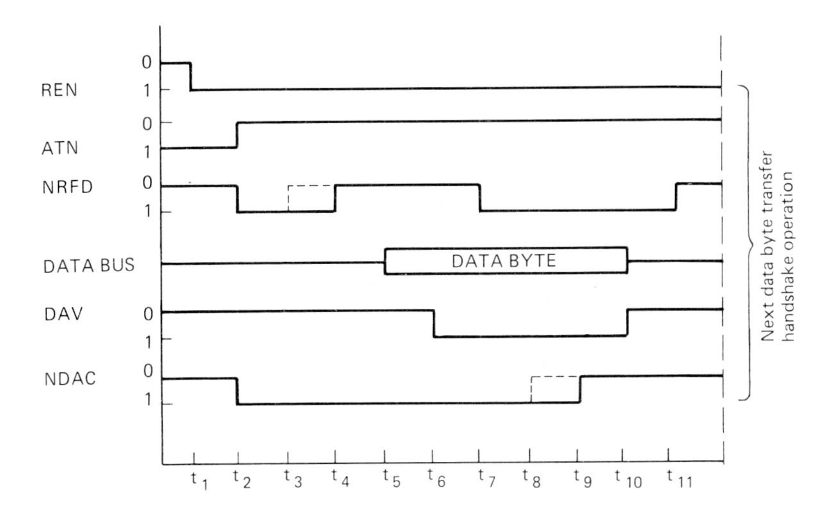

Figure 11.7 Timing diagram of signals on the GPIB, illustrating the handshake procedure between talkers and listeners that is required to allow transmission of each data byte across the data bus

logic level only changes when *all* listener instruments are ready to accept data (NRFD at t_4).

The talker now presents the data byte on lines DIO 1 to DIO 8, that is, the data bus (at t_5); and then indicates this with a *data valid* (DAV at t_6) signal. The listeners reset the *not ready for data* (NRFD at t_7) signal.

At t_8 the fastest instrument has accepted the data byte, but the *not data accepted* (NDAC) signal line is only reset when *all listeners have accepted the data byte, at* t_9.

The talker now removes the data byte and clears the *data valid* (DAV) signal, at t_{10}. After an interval, which depends on the time each listening instrument needs to process the data byte, the *not ready for data* (NRFD at t_{11}) signal changes once again and the data transfer proceeds.

As three signal lines are used for this handshake – *data valid* (DAV), *not ready for data* (NRFD) and *not data accepted* (NDAC) – the procedure is often called a **three-line handshake**. It provides a simple but effective method of transferring data bytes over the GPIB, with the advantage that it doesn't take up much of the controller's processing time.

Limitations

The GPIB has a number of limitations which should be borne in mind when constructing an automatic test equipment system. First, the maximum number of instruments (including the controller) that can be connected to the bus is 15. The maximum cable length is calculated as:

2 metres times the number of instruments

or:

20 metres

whichever is less. The maximum distance between any two instruments is 4 metres.

With a 20 metre cable, the bus can operate with a data transfer rate of up to 250 000 bytes s^{-1}, but with a shorter cable length considerably higher data transfer rates (up to 1 000 000 bytes s^{-1}) can be achieved. In most practical applications, however, overall data transfer rate is much less than these maxima, being limited by the speed of operation of the instruments connected.

Table 11.11 lists pin assignments for the GPIB,

Table 11.11 GPIB connector pin assignments

Pin no.	*Signal group*	*Abbreviation*	*Signal/function*
1	Data	D101	Data line 1
2		D102	Data line 2
3		D103	Data line 3
4		D104	Data line 4
5	Management	EOI	End or identify (sent by a talker to indicate that transfer of data is complete)

Pin no.	*Signal group*	*Abbreviation*	*Signal/function*
6	Handshake	DAV	Data valid (asserted by a talker to indicate that valid data is present on the bus)
7		NRFD	Not ready for data (asserted by a listener to indicate that it is not ready for data)
8		NDAC	Not data accepted (asserted while data is being accepted by a listener)
9	Management	IFC	Interface clear (asserted by the controller in order to initialise the system in known state)
10		SRQ	Service request (sent to the controller by a device requiring attention)
11		ATN	Attention (asserted by the controller when placing a command onto the bus)
12		SHIELD	Shield
13	Data	D105	Data line 5
14		D106	Data line 6
15		D107	Data line 7
16		D108	Data line 8
17	Management	REN	Remote enable (enables an instrument to be controlled by the bus controller rather than by its own front panel controls)
18		GND	Ground/common
19		GND	Ground/common
20		GND	Ground/common
21		GND	Ground/common
22		GND	Ground/common
23		GND	Ground/common
24		GND	Ground/common

Notes: (a) Handshake signals (DAV, NRFD and NDAC) are all active low open collector and are used in a wired-OR configuration.

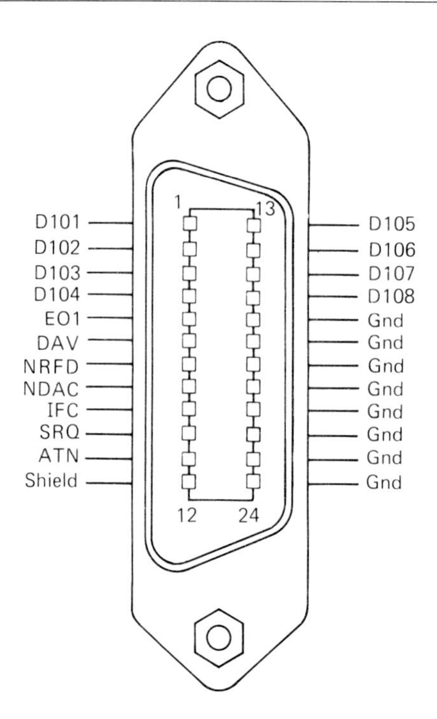

Figure 11.8 GPIB connector pin assignments

while Figure 11.8 illustrates pin numbering of the connector.

VXIbus

Physical structure of a VXIbus automatic test equipment system comprises a mainframe chassis with backplane and connections to allow modules to be plugged in as required (Figure 11.9). It is thus a classical modular-based backplaned computer system. However, logical design is structured specifically with an automatic test equipment system in mind. Consequently, a VXIbus automatic test equipment system has the benefits of high data rate associated with a modular-based backplaned computer system, coupled with the ease-of-use and controllability of a purpose-built automatic test equipment system. Modules are inserted into slots in the backplaned

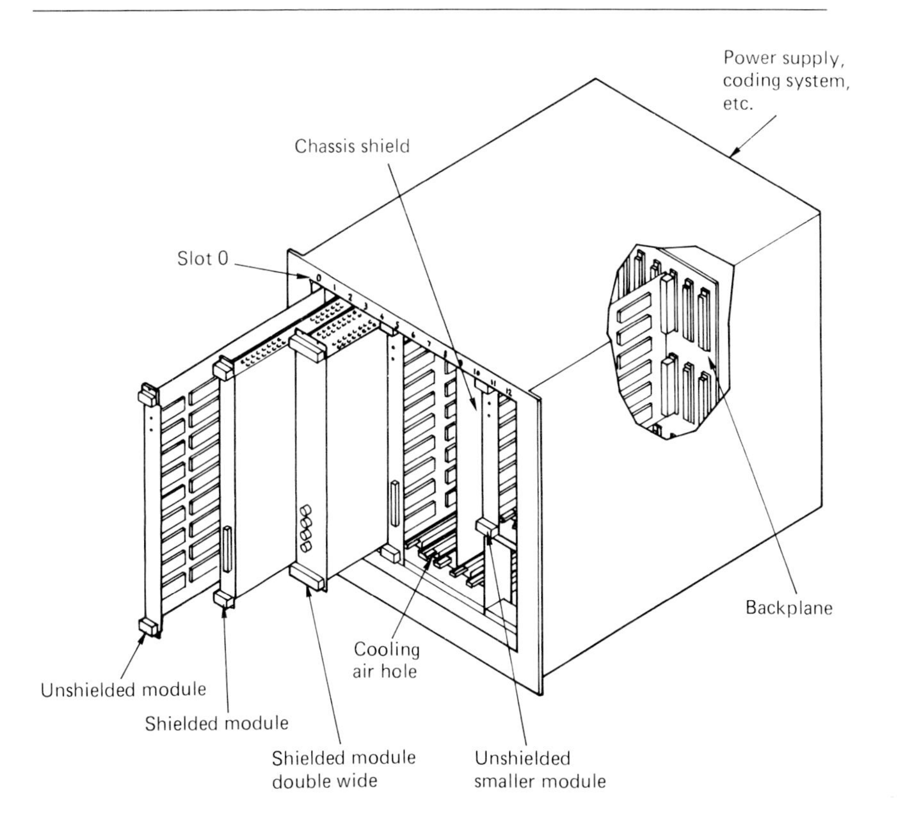

Figure 11.9 Classical, modular-based backplaned bus system used in VXIbus

housing. Slot 0 is always occupied by a timing and control module.

An individual test instrument for use on VXIbus is known somewhat ambiguously as a **device**. A single module on VXIbus may house one or more devices. On the other hand, a complex device may be formed by one or more modules. Both these extremes are catered for by the logical structure of VXIbus.

Up to thirteen modules (one of which is a timing module) may be housed in a single housing called a **subsystem**. If a larger system is required, individual subsystems may be connected, to a maximum of 255 devices.

Logical structure of VXIbus is totally open. That is,

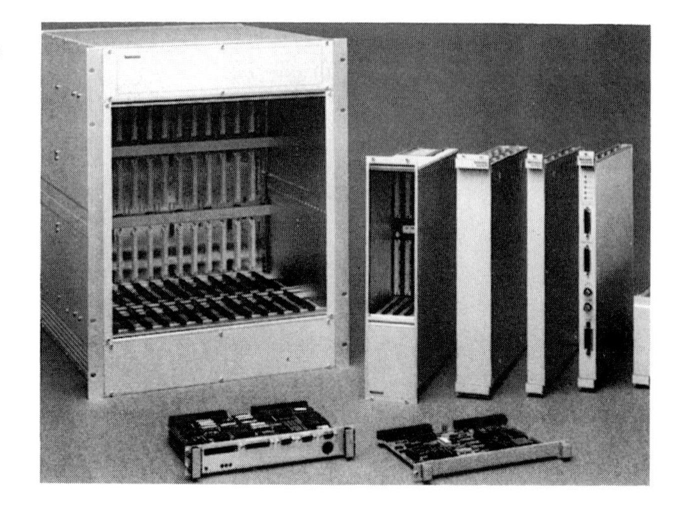

Photo 11.1 Typical VXIbus automatic test equipment bus system (Tektronix)

Photo 11.2 VXIbus controller, keyboard and screen (Tektronix)

operating system, microprocessor, computer interface, hierarchy and so on are all undefined. Only protocols and criteria which must exist to ensure compatibility between modules are defined, and these do not interfere with individual device microprocessor controllers.

VXIbus is based on the VMEbus (in fact, VXIbus stands for *VME extensions for instrumentation bus*) and uses the complete VMEbus backplaned structure using P1 and P2 connectors between modules and backplane, adding to it in a number of ways. Mechanical differences include:

- slightly wider spacing between boards, to allow VXIbus modules to be shielded in metal cases – this means that VMEbus modules can fit into a VXIbus chassis, but not vice versa
- a complete specification of packaging requirements for modules; electromagnetic compatibility, power distribution, cooling and air flow in the chassis, and so on
- two new module sizes, C- and D-sized modules, with a further optional connector P3 on the largest D-sized module. Table 11.12 lists all VXIbus module sizes and connectors.

Table 11.12 VXIbus module sizes and connectors

Module	*Connectors*	*Sizes (mm)*
A	P1	99 by 160 by 20
B	P1 P2 (optional)	233 by 160 by 20
C	P1 P2 (optional)	233 by 340 by 30
D	P1 P2, P3 (both optional)	365 by 240 by 30

There are significant logical differences, too, including:

- definition of remaining pin assignments on the P_2 connector, and all P_3 connector assignments
- possibilities of low level and high level communications between devices on the bus.

Devices and communications

Communications needs of all devices within a VXIbus system is catered for with a layered set of communications protocols, as illustrated in Figure 11.10.

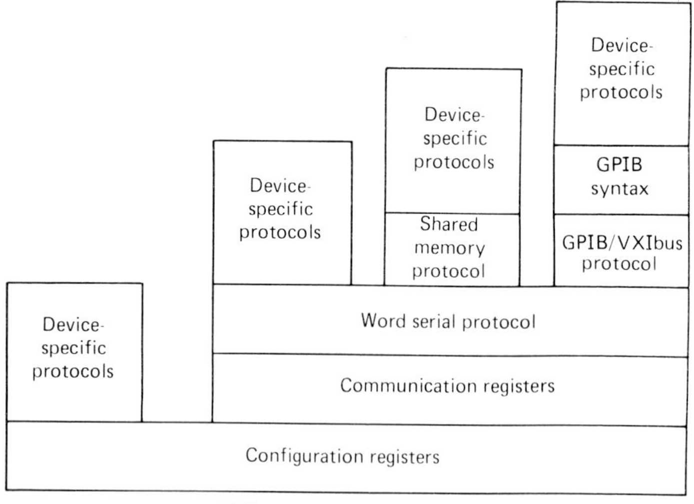

Figure 11.10 VXIbus communication protocol layers

Each device has a set of **configuration registers**, containing information relating to the device's logical address, address space and memory requirements, type, model and manufacturer (Figure 11.11). These configuration registers guarantee automatic system and memory configuration for each device, on application of power, at least to a minimum degree

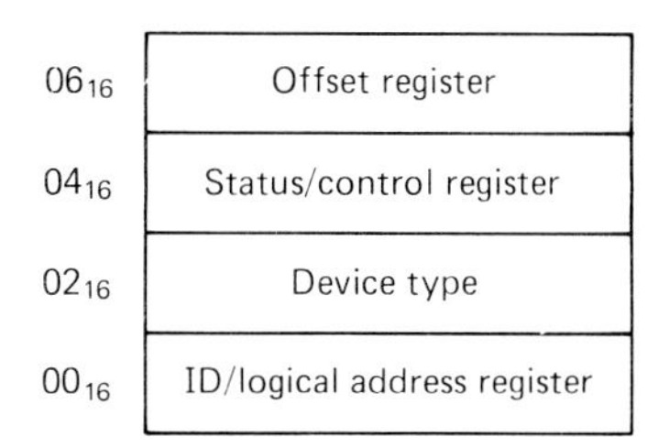

Figure 11.11 Register-based VXIbus device configuration register

of communication. Devices with this minimum configuration level alone are known as **register-based devices**.

A further set of registers, known as **communication registers** (Figure 11.12), is included if a device is to

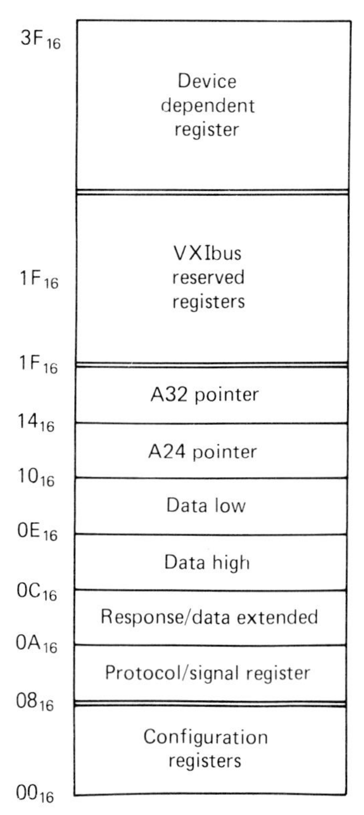

Figure 11.12 Message-based VXIbus device communication registers

communicate at a higher level with other devices. Devices with communication registers are known as **message-based devices** and are all able to communicate, initially at least, with a specific protocol called **word serial protocol**. Once communication is established using word serial protocol, devices may opt to progress to higher performance protocols. By no means a coincidence, VXIbus word serial protocol is very similar to GPIB communications protocol. One benefit of agreement by manufacturers to build devices which use standardised communications protocols such as word serial protocol is that devices from any manufacturer will be compatible. Higher level protocols may be defined later, on agreement by manufacturers, perhaps to be standardised into the VXIbus system specification.

Devices may be of purely memory form, say, RAM or ROM. Configuration registers of such **memory devices** are illustrated in Figure 11.13.

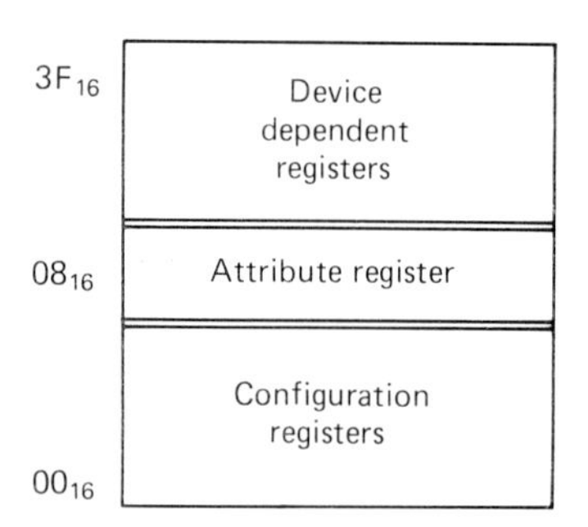

Figure 11.13 VXIbus memory device configuration register

In some systems, **extended devices** are used which allow for definition of sub-classes of device types not defined in the original specification. Configuration registers of extended devices are illustrated in Figure 11.14, where the sub-class register allows for definition of both standard and manufacturer-specific extended device sub-classes.

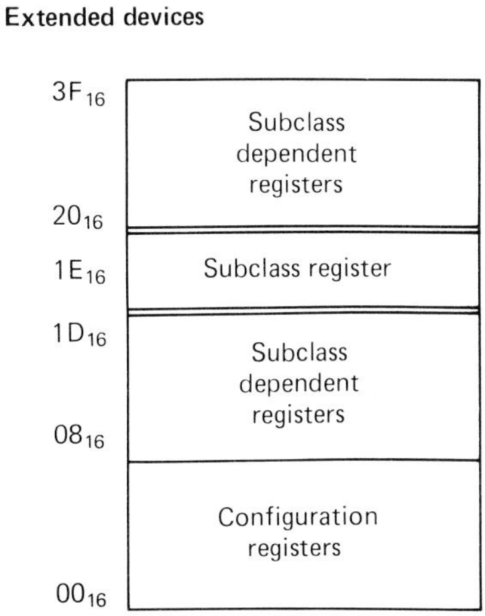

Extended devices

Address	Register
$3F_{16}$ – 20_{16}	Subclass dependent registers
$1E_{16}$	Subclass register
$1D_{16}$ – 08_{16}	Subclass dependent registers
00_{16}	Configuration registers

Figure 11.14 VXIbus extended device configuration register

These make up the four classes of VXIbus devices. In addition, there are two classes of VMEbus device which may be used in a VXIbus system. First, is the **hybrid device**; a VMEbus compatible device which has the ability to communicate with VXIbus devices, but does not comply with VXIbus device requirements. Second, is the **non-VXIbus device**; a VMEbus device not using or complying with any VXIbus requirement.

VXIbus devices of all classes fall into a hierarchical structure, as illustrated in Figure 11.15.

Control

Control of the logical resources (memory, diagnostic procedures, self-test analysis and so on) of a VXIbus system is exercised by a **resource manager** device, located in slot 0 and given the logical address 0.

On application of power it is the resource manager's function to read the logical addresses of each device in the system, thereby locating all device configuration registers. Once these have been read the resource manager knows total system requirements, and allocates resources to suit.

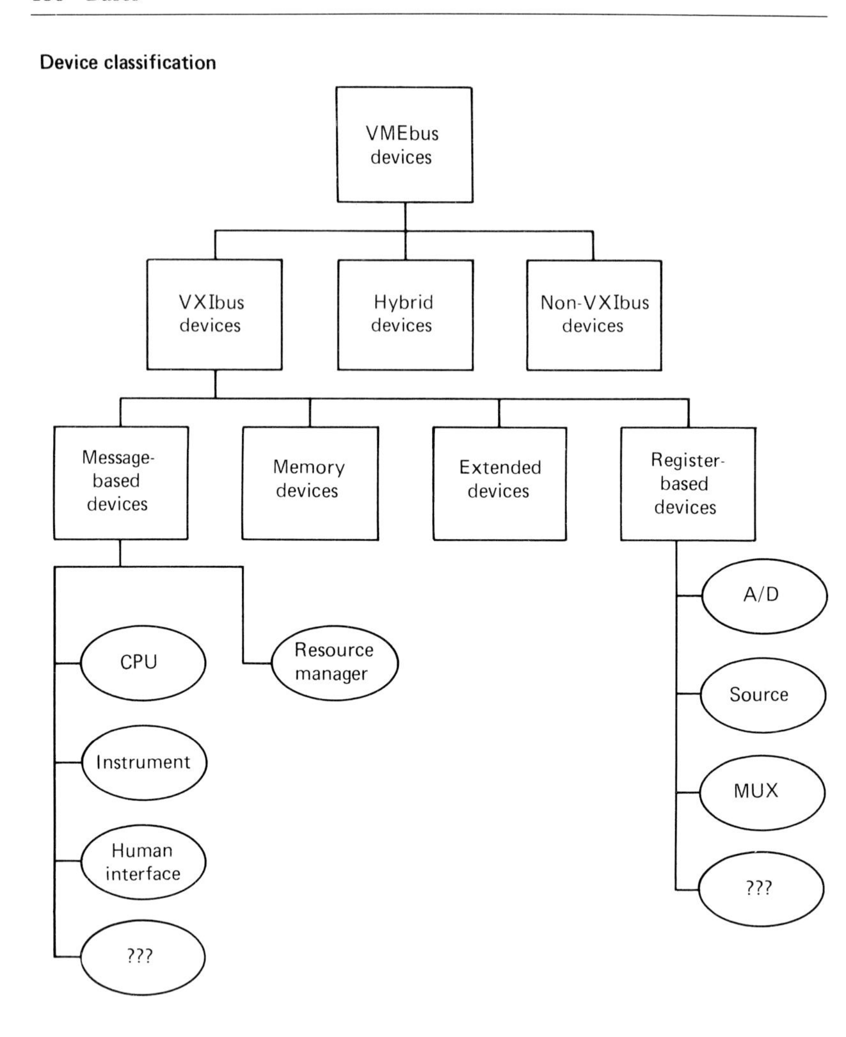

Figure 11.15 VXIbus device hierarchical classifications

Where systems feature more than one microprocessor-controlled device the resource manager also determines hierarchies between them, such that devices are allocated as **commanders** or **servants**.

Often resource manager functions in slot 0 are coupled with general-purpose devices such as a VXIbus-to-GPIB interface, printer output and general timing functions.

Overall structure for a VXIbus system with P2 connector is:

- the VMEbus
- 13 module identity (MODID) lines
- 12 local bus (LBUS) lines
- 8 TTL trigger (TTLTR) lines
- 2 ECL trigger (ECLTR) lines
- 1 analog summing bus (SUMBUS) line – terminated by 50 Ω
- 2 clock (CLK10) lines – at 10 MHz
- 2 lines reserved for future use.

Overall structure for a VXIbus system with P3 connector is:

- VXIbus system with P2 connector
- 24 additional local bus lines
- 4 additional ECL trigger lines
- 52 star trigger (STAR) lines for precision module-to-module timing
- 2 clock (CLK100) lines – at 100 MHz (synchronous with P2 CLK10)
- 2 synchronising signal (SYNC100) lines
- 2 additional power supply lines
- 4 lines reserved for future use.

Table 11.13 lists pin assignments for the fully defined P2 connector in slot 0, while Table 11.14 lists same for other slots. Similarly, Table 11.15 lists pin assignments for the P3 connector in slot 0, while Table 11.16 lists same for other slots.

Readers may wish to consider VXIbus in greater depth than is possible here. If so, refer to:

- report 88–0529R: *The VXIbus* by ERA Technology Ltd, Cleeve Road, Leatherhead, Surrey KT22 7SA. This gives a concise description of VXIbus,

Table 11.13 VXIbus slot 0 P2 connector pin assignments

Pin Number	*Row a signal mnemonic*	*Row b signal mnemonic*	*Row c signal mnemonic*	*Pin number*
1	ECLTRG0	+5 V	CLK10+	1
2	−2 V	GND	CLK10−	2
3	ECLT RG1	RSV1	GND	3
4	GND	A24	−5.2 V	4
5	MODID12	A25	LEBUSC00	5
6	MODID11	A26	LBUSC01	6
7	−5.2 V	A27	GND	7
8	MODID10	A28	LBUSC02	8
9	MODID09	A29	LBUSC03	9
10	GND	A30	GND	10
11	MODID08	A31	LBUSC04	11
12	MODID07	GND	LBUSC05	12
13	−5.2 V	+5 V	−2 V	13
14	MJODID06	D16	LBUSC06	14
15	MODID05	D17	LBUSC07	15
16	GND	D18	GND	16
17	MODID04	D19	LBUSC08	17
18	MODID03	D20	LBUSC09	18
19	−5.2 V	D21	−5.2 V	19
20	MODID02	D22	LBUSC10	20
21	MODID01	D23	LBUSC11	21
22	GND	GND	GND	22
23	TTLTRG0*	D24	TTLTRG1*	23
24	TTLTRG2*	D25	TTLTRG3*	23
25	+5 V	D26	GND	25
26	TTLTRG4*	D27	TTLTRG5*	26
27	TTLTRG6*	D28	TTLTRG7*	27
28	GND	D29	GND	28
29	RSV2	D30	RSV3	29
30	MODID00	D31	GND	30
31	GND	GND	+24 V	31
32	SUMBUS	+5 V	−24 V	32

Table 11.14 VXIbus slots 1 to 12 P2 connector pin assignments

Pin Number	*Row a signal mnemonic*	*Row b signal mnemonic*	*Row c signal mnemonic*	*Pin number*
1	ECLTRG0	+5 V	CLK10+	1
2	−2 V	GND	CLK10−	2
3	ECLTRG1	RSV1	GND	3
4	GND	A24	−5.2 V	4
5	LBUSA00	A25	LEBUSC00	5
6	LBUSA01	A26	LBUSC01	6
7	−5.2 V	A27	GND	7
8	LBUSA02	A28	LBUSC02	8
9	LBUSA03	A29	LBUSC03	9
10	GND	A30	GND	10
11	LBUSA04	A31	LBUSC04	11
12	LBUSA05	GND	LBUSC05	12
13	−5.2 V	+5 V	−2 V	13
14	LBUSA06	D16	LBUSC06	14
15	LBUSA07	D17	LBUSC07	15
16	GND	D18	GND	16
17	LBUSA08	D19	LBUSC08	17
18	LBUSA09	D20	LBUSC09	18
19	−5.2 V	D21	−5.2 V	19
20	LBUSA10	D22	LBUSC10	20
21	LBUSA11	D23	LBUSC11	21
22	GND	GND	GND	22
23	TTLTRG0*	D24	TTLTRG1*	23
24	TTLTRG2*	D25	TTLTRG3*	23
25	+5 V	D26	GND	25
26	TTLTRG4*	D27	TTLTRG5*	26
27	TTLTRG6*	D28	TTLTRG7*	27
28	GND	D29	GND	28
29	RSV2	D30	RSV3	29
30	MODID	D31	GND	30
31	GND	GND	+24 V	31
32	SUMBUS	+5 V	−24 V	32

Table 11.15 VXIbus slot 0 P3 connector pin assignments

Pin Number	*Row a signal mnemonic*	*Row b signal mnemonic*	*Row c signal mnemonic*	*Pin number*
1	ECLTRG2	+24 V	+12 V	1
2	GND	−24 V	−12 V	2
3	ECLTRG3	GND	RSV4	3
4	−2 V	RSV5	+5 V	4
5	ECLTRG4	−5.2 V	RSV6	5
6	GND	RSV7	GND	6
7	ECLTRG5	+5 V	−5.2 V	7
8	−2 V	GND	GND	8
9	STARY12+	+5 V	STARX01+	9
10	STARY12−	STARY01−	STARX01−	10
11	STARX12+	STARX12−	STARY01+	11
12	STARY11+	GND	STARX02+	12
13	STARY11−	STARY02−	STARX02−	13
14	STARX11+	STARX11−	STARY02+	14
15	STARY10+	+5 V	STARX03+	15
16	STARY10−	STARY03−	STARX03−	16
17	STARX10+	STARX10−	STARY03+	17
18	STARY09+	−2 V	STARX04+	18
19	STARY09−	STARY04−	STARX04−	19
20	STARX09+	STARX09−	STARY04+	20
21	STARY08+	GND	STARX05+	21
22	STARY08−	STARY05−	STARX05−	22
23	STARX08+	STARX08−	STARY05+	23
24	STARY07+	+5 V	STARX06+	24
25	STARY07−	STARY06−	STARX06−	25
26	STARX07+	STARX07−	STARY06+	26
27	GND	GND	GND	27
28	STARX+	−5.2 V	STARY+	28
29	STARX−	GND	STARY−	29
30	GND	−5.2 V	−5.2 V	30
31	CLK100+	−2 V	SYNC100+	31
32	CLK100−	GND	SYNC100−	32

Table 11.16 VXIbus slots 1 to 12 P3 connector pin assignments

Pin Number	*Row a signal mnemonic*	*Row b signal mnemonic*	*Row c signal mnemonic*	*Pin number*
1	ECLTRG2	+24 V	+12 V	1
2	GND	−24 V	−12 V	2
3	ECLTRG3	GND	RSV4	3
4	−2 V	RSV5	+5 V	4
5	ECLTRG4	−5.2 V	RSV6	5
6	GND	RSV7	GND	6
7	ECLTRG5	+5 V	−5.2 V	7
8	−2 V	GND	GND	8
9	LBUSA12	+5 V	LBUSC12	9
10	LBUSA13	LBUSC15	LBUSC13	10
11	LBUSA14	LBUSA15	LBUSC14	11
12	LBUSA16	GND	LBUSC16	12
13	LBUSA17	LBUSC19	LBUSC17	13
14	LBUSA18	LBUSA19	LBUSC18	14
15	LBUSA20	+5 V	LBUSC20	15
16	LBUSA21	LBUSC23	LBUSC21	16
17	LBUSA22	LBUSA23	LBUSC22	17
18	LBUSA24	−2 V	LBUSC24	18
19	LBUSA25	LBUSC27	LBUSC25	19
20	LBUSA26	LBUSA27	LBUSC26	20
21	LBUSA28	GND	LBUSC28	21
22	LBUSA29	LBUSC31	LBUSC29	22
23	LBUSA30	LBUSA31	LBUSC30	23
24	LBUSA32	+5 V	LBUSC32	24
25	LBUSA33	LBUSC35	LBUSC33	25
26	LBUSA34	LBUSA35	LBUSC34	26
27	GND	GND	GND	27
28	STARX+	−5.2 V	STARY+	28
29	STARX−	GND	STARY−	29
30	GND	−5.2 V	−5.2 V	30
31	CLK100+	−2 V	SYNC100+	31
32	CLK100−	GND	SYNC100−	32

together with general descriptions of equipment and instruments available

- *VXIbus system specification*, generally available from VXIbus device manufacturers. This is the in-depth specification of the VXIbus, detailing signals and timing considerations.

Part Two Applications and Measurement Techniques

12 Measurands and measurement equipment

What follows is a discussion of various measurands which are measured, monitored, or tested by electronic test equipment. The discussion is as exhaustive as possible. Measurands are in a simple alphabetic order, followed by the methods available.

Where more than one method may be used to measure the measurand, these are then listed alphabetically, too, according to the equipment.

In many cases measurement of a measurand is undertaken simply by using a transducer which converts the measurand into an electrical quantity (voltage, current and so on) which is measured using fairly basic equipment. For this reason the following discussion tends to discuss measurements in terms of these basic electrical quantities (in themselves, measurands) although certain important measurands which are not basic are also covered.

Aerials or antennas

Electronic test equipment is used to measure a number of aerial parameters.

Directivity

Aerial directivity is measured with a field strength meter, by rotating the aerial while it is being used for transmission, and measuring the strength of the transmitted signal.

Gain

Aerial gain is usually quoted for a receiving aerial, and is a direct comparison of the received signal from the aerial with that from a dipole of the same size. A wavemeter is used to measure received signals from both aerials and the gain is expressed in decibels relative to the dipole (dBd). Care should be taken to ensure identical aerial conditions in the two measurements.

Impedance, radiation resistance
Although aerial impedance is an important factor, radiation resistance of an aerial is more meaningful, usually measured with an ammeter and signal source, as shown in Figure 12.1. Here variable capacitor VC and variable resistor VR are used to form an equivalent circuit to the aerial. Initially, with switch SW feeding the aerial, the aerial is tuned to the RF signal from the signal source by adjusting variable inductor VL till the ammeter reads a maximum value, *I*.

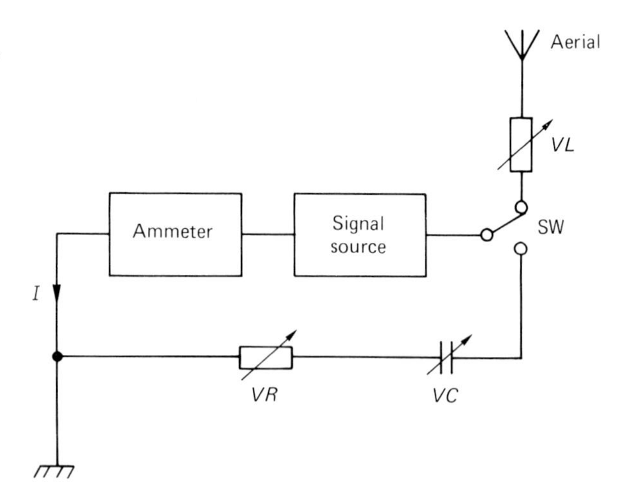

Figure 12.1 Measurement of aerial radiation resistance

Next, the variable capacitor and resistor are switched into circuit and the variable capacitor is adjusted till the ammeter is another maximum. Finally, the variable resistor is adjusted till the ammeter displays the value *I*. Value of the variable

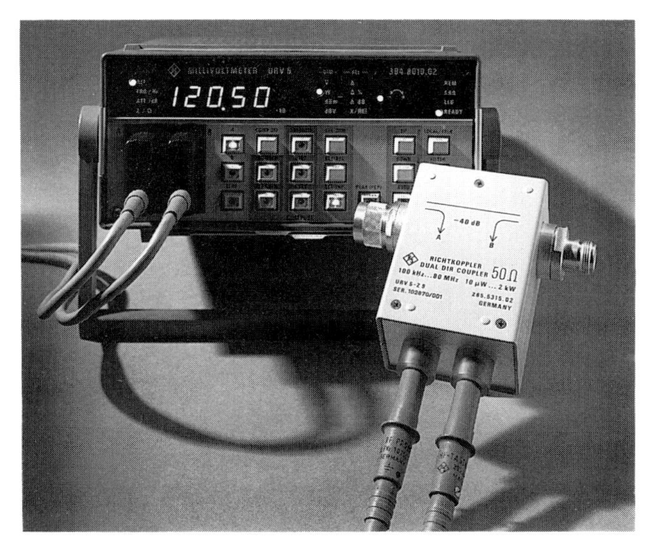

Photo 12.1 Rohde & Schwarz dual directional coupler used with millivoltmeter to measure forward and reflected power, reflection coefficient and return loss (Rohde & Schwarz)

resistance is now equal to the radiation resistance of the aerial at the frequency in the test.

Resonant frequency

There are a number of methods commonly used to measure an aerial's resonant frequency:

- the *series ammeter* method, shown in Figure 12.2, where a signal source is tuned to give maximum

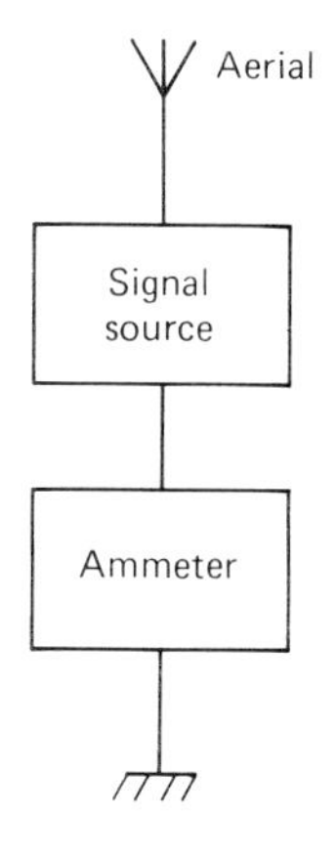

Figure 12.2 Measurement of aerial resonant frequency by the series ammeter method

reading on a series connected ammeter. At this reading the maximum transfer of energy occurs between signal source and aerial, and the resonant frequency is simply read from the signal source

- using a grid dip meter, coupled to the aerial by its tuning coil, and a signal source, as shown in Figure 12.3. Sweeping the signal source from the grid dip meter will cause the aerial's resonant frequency to be indicated where the meter reading dips. The lowest frequency dip corresponds to the aerial's primary resonant frequency, while higher frequency dips are harmonics, at exact multiples

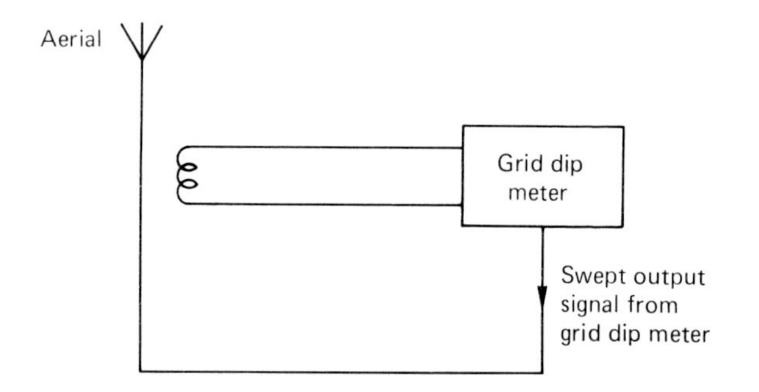

Figure 12.3 Using a grid dip meter to measure an aerial's resonant frequency

- using an oscilloscope and sweep generator signal source, a display of aerial transmitted signal against applied frequency may be achieved. The setup shown in Figure 12.4 illustrates the sweep generator applied to the aerial and the oscilloscope's horizontal input, while a pick-up monitors the aerial output and applies this to the oscilloscope's vertical input
- a wavemeter can be used, as shown in Figure 12.5. By adjusting the internal tuned circuit to give a maximum reading on the wavemeter, the aerial's resonant frequency is found.

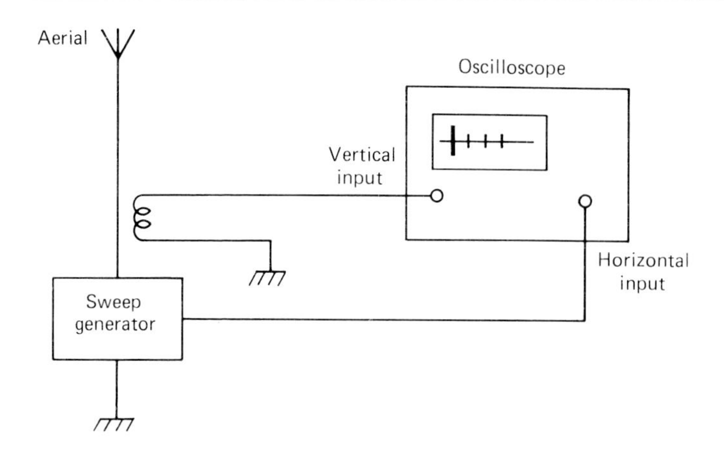

Figure 12.4 Using an oscilloscope to display the aerial's transmitted signal against frequency, allowing resonant frequency to be measured

Transmitted power

Using the procedure earlier to measure aerial radiation resistance, the aerial's radiated power is given by the relationship:

$$P = I^2R$$

If a transmitting aerial is to be setup for maximum power output, a wavemeter should be tuned to the transmitter's carrier frequency, as illustrated in Figure 12.6. The tuning component (either a variable capacitor or inductor) on the transmitting aerial

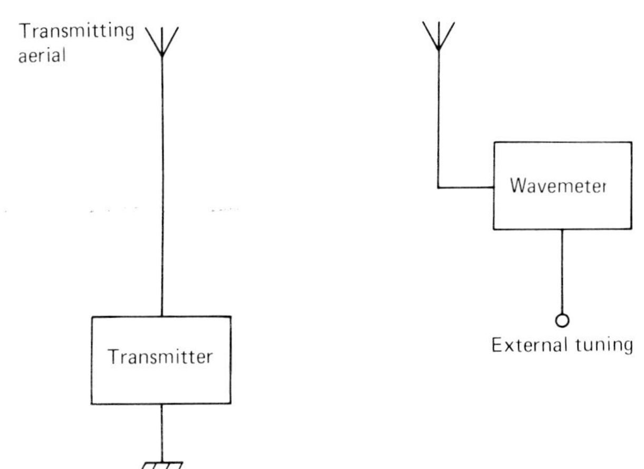

Figure 12.5 Measurement of aerial resonant frequency using a wavemeter

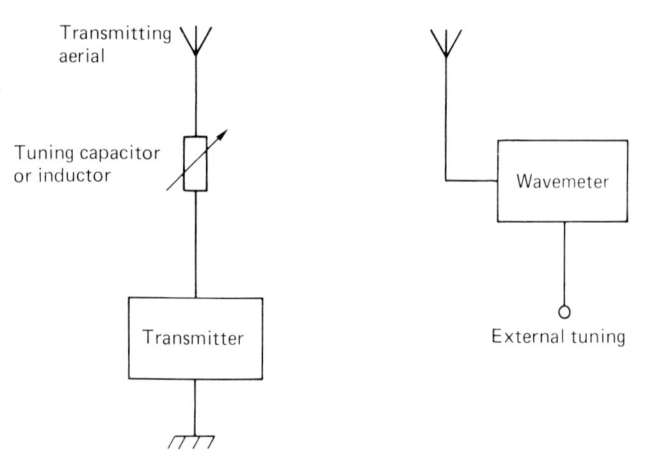

Figure 12.6 A wavemeter can be used to aid setting up of an aerial for maximum radiated power

should then be adjusted for maximum wavemeter reading.

Amplifiers (untuned)

There are two main types of untuned amplifier: audio and video. Audio amplifiers and circuits operate to amplify or adjust signals within the audio hearing range (possibly just outside the limits, too). Video amplifiers and circuits do similar with signals over a wider band of frequencies (say, from DC to many megahertz). Many measurements may be required on untuned amplifiers and circuits, and the following examples include some of the more common. In the majority of cases these measurements are made with relatively simple and common test equipment.

Common mode rejection ratio

Operational amplifier common mode rejection ratio is measured by adjusting the value of the differential input voltage V_{diff} to obtain a conveniently measurable output voltage V_{out}, as shown in Figure 12.7a. The differential inputs should then be shorted and the common mode input voltage V_{comm}, adjusted till the output voltage V_{out} is the same as before (Figure 12.7b). Common mode rejection ratio (CMRR) is then given by:

$$CMRR = V_{comm}/V_{diff}$$

Frequency response or bandwidth

Frequency response or bandwidth of a circuit defines the range of signal frequencies which the circuit passes. Generally, the frequency response is given as two frequencies at each end of the frequency range where the amplitude of the output signal has fallen by a specified amount from the amplitude in the middle band of frequencies. The specified amount is usually taken to be where the output signal power falls to a half of the mid-band power, equivalent to

Figure 12.7 Operational amplifier common mode rejection ratio (a) measuring output voltage V_{out} for a particular differential input voltage V_{diff} (b) determining the same output voltage V_{out} for a common mode input voltage V_{comm}

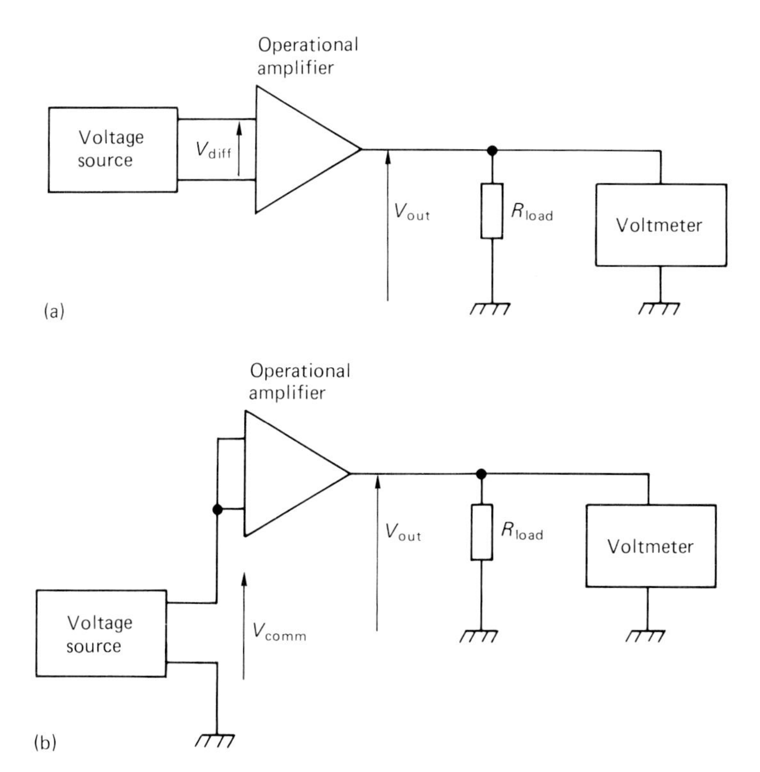

an output signal voltage of about 0.7 of the mid-band signal voltage. There are three main ways in which frequency response is measured:

- using a signal source and AC voltmeter as shown in Figure 12.8a, the output amplitude is monitored as the input signal frequency is swept. Care must be taken to ensure the input signal amplitude does not vary (or at least is taken into account when measuring the output signal amplitude) and so an AC voltmeter is often also included at the amplifier input.
- using a power meter and signal source. Power readings are taken over the range, and where the power falls to one half of the mid-band range (or −3 dB, if calibrated in decibels) defines the frequency response limits.
- using a spectrum analyser, as shown in Figure 12.8b, a graphic display of frequency response is obtained, allowing the upper and lower frequencies to be measured.

The first two are time domain measurements, while the third is in the frequency domain.

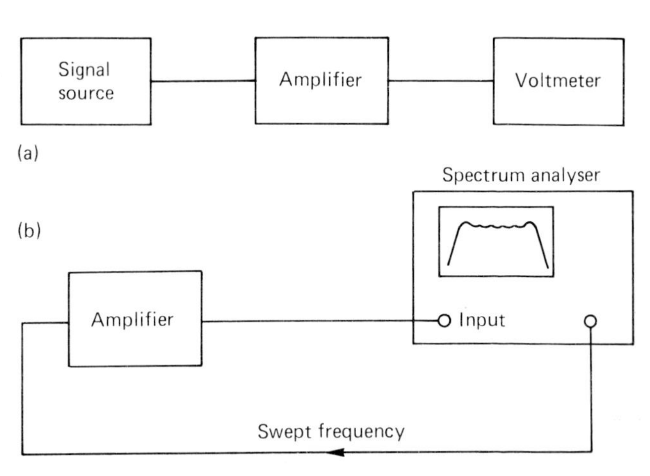

Figure 12.8 Measuring analog circuit frequency response or bandwidth, using (a) signal source and voltmeter (b) spectrum analyser

Gain

A circuit's voltage gain may be measured indirectly, using the setups used to measure frequency response, and is calculated from the ratio V_{out}/V_{in}. Power gain, on the other hand, is the ratio of output power P_{out} to input power P_{in}. Measurement of amplifier powers is described later.

Impedance

Setup of Figure 12.7 may be used to measure dynamic output impedance, and at any particular frequency the load resistance is changed until output power is a maximum. Value of the load resistance is equal to the dynamic impedance.

Figure 12.9 shows a configuration enabling the dynamic input impedance of an amplifier to be measured. Initially, at a particular frequency, a voltmeter is used to measure the voltage across a series resistor V_{vr}. Then it is used to measure the amplifier's input voltage V_{in}. By alternating these measurements (or using two voltmeters) and adjusting the value of VR until the voltages are equal, a potential divider is formed in which the resistance of resistor VR equals the input impedance of the amplifier.

Power

Output power can be calculated from the output

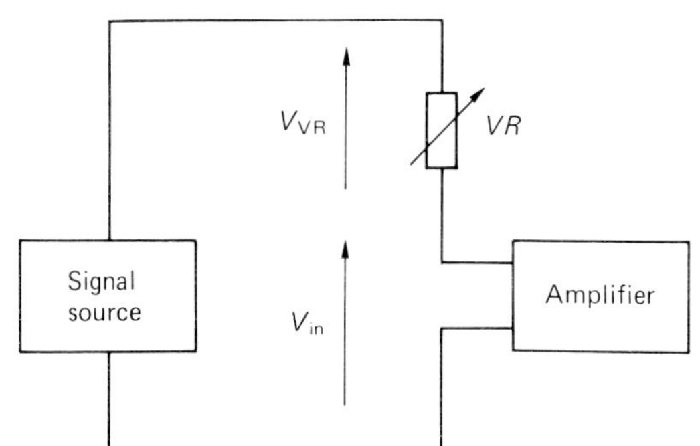

Figure 12.9 Dynamic input impedance of an amplifier may be determined by measuring voltages across the amplifier's input and a series resistance

voltage (measured with a voltmeter) and load resistance according to the relationship:

$$P_{out} = V_{out}^2 / R_{load}$$

Alternatively, many types of power meter exist to directly measure output power.

Input power must be calculated by first measuring the input impedance and input voltage, and is given by the relationship:

$$P_{in} = V_{in}^2 / R_{in}$$

Sensitivity

Input sensitivity is the ratio of output power to input voltage P_{out}/V_{in}, measurements of which are described earlier.

Slew rate

Figure 12.10 illustrates a configuration which can be used to measure amplifier slew rate, which is the ratio of the rate of change of output voltage to the rate of change of time, $\Delta V_{out}/\Delta t$. Usually, a unity gain

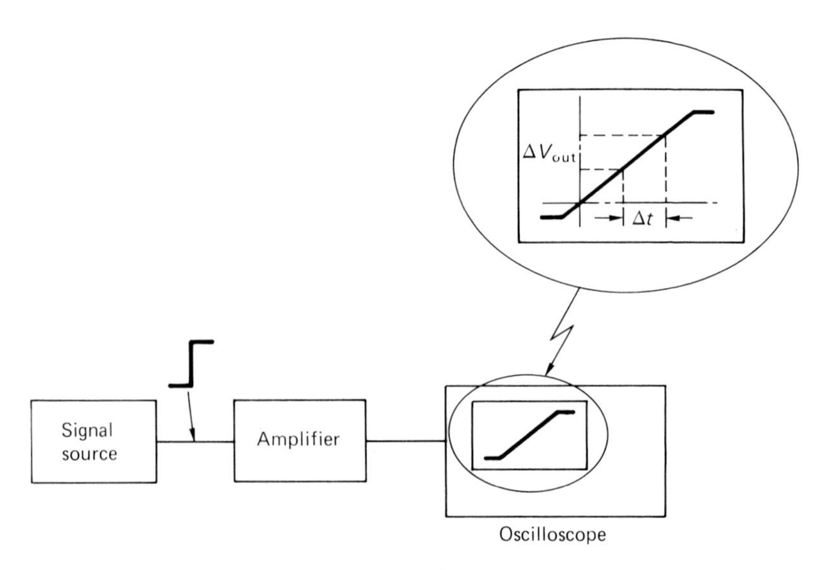

Figure 12.10 Determining an amplifier's slew rate by measuring the rate of change of amplifier output voltage and rate of change of time on an oscilloscope

non-inverting follower circuit is tested, which gives worst-case results. Requirements of the test are simply to cause a reversal of saturated output, so a simple pulse input at the correct amplitude will suffice. Either a storage oscilloscope of some description, or a real-time oscilloscope and a squarewave input, will allow measurement.

More complicated test equipment is also often used to measure untuned amplifier parameters. For audio work, network analysers are common; while video circuit parameters are sometimes measured with vector voltmeters.

Capacitance

There are quite a number of methods of measuring capacitance, including:

- bridge circuits – many types are used. Perhaps one of the most convenient, as it trebles as a bridge to measure inductance and resistance, is the *substitution bridge* of which, again, there are a number of types. Figure 12.11 shows a common version.

  Initially, points 1 and 2 are linked and the bridge is adjusted for balance by means of

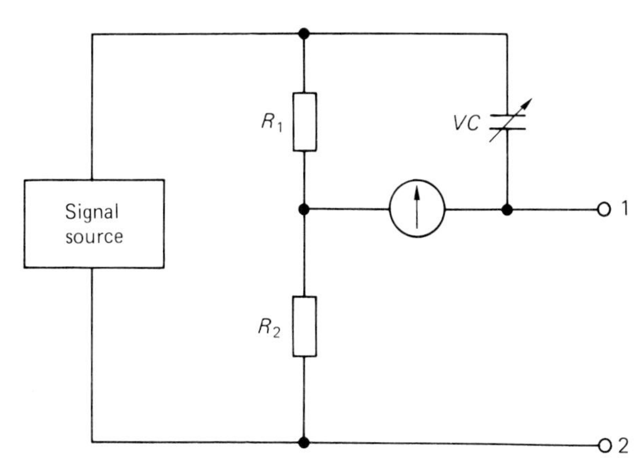

Figure 12.11 Using a substitution bridge to measure an unknown capacitor value

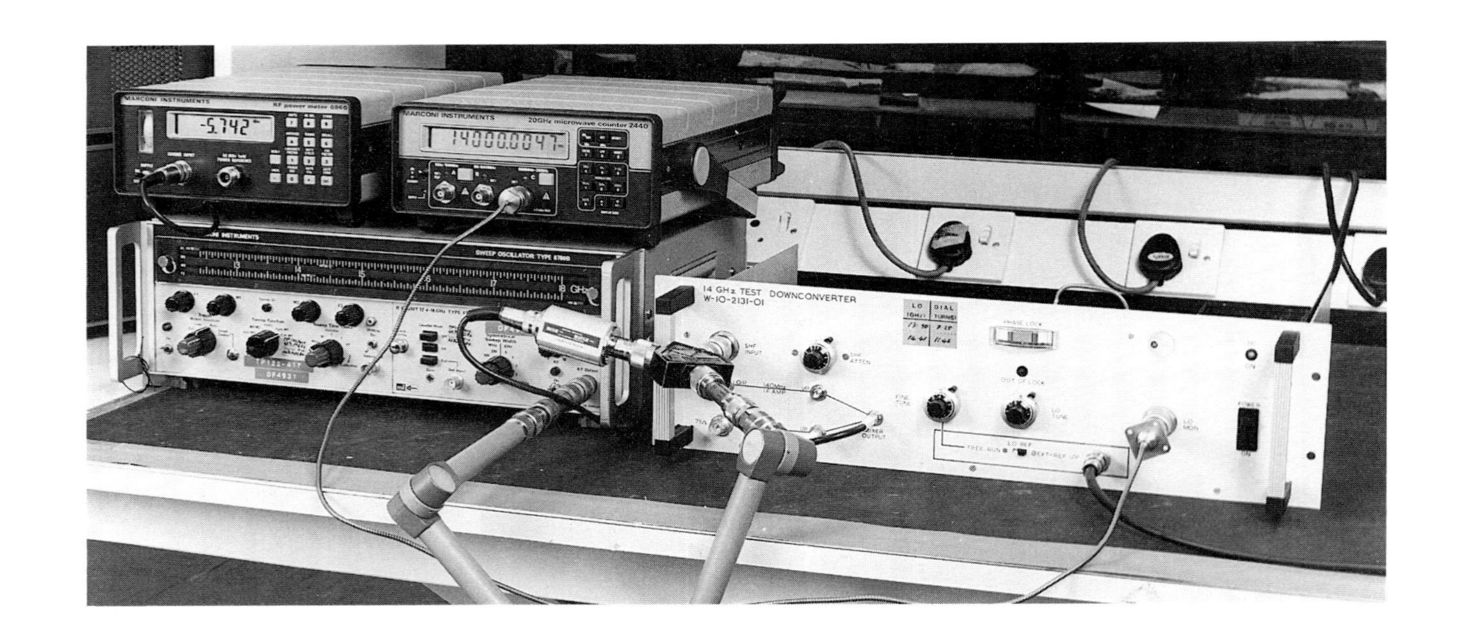
MARCONI INSTRUMENTS
RF power meter 6960
-5.742
MARCONI INSTRUMENTS
20GHz microwave counter 2440
14000.0047
SWEEP OSCILLATOR TYPE 6700B
13
14
15
16
17
18 GHz
14 GHz TEST DOWNCONVERTER
W-10-2131-01
SHF INPUT
SHF ATTEN
PHASE LOCK
OUT OF LOCK
FINE TUNE
LO TUNE
LO MON
MIXER OUTPUT
LO REF
FREE RUN
POWER
ON

variable capacitor VC. The value of the variable CVC_a is noted, and the unknown capacitor C_u replaces the link between points 1 and 2. Once again the bridge is balanced, with the new value for the variable CVC_b noted. The value of the unknown capacitor is calculated from the expression:

$$C_u = \frac{R_1\, CVC_a}{R_2}\left(\frac{CVC_a}{CVC_a - CVC_b} - 1\right)$$

- a counting circuit, such as that shown in Figure 12.12. Here a monostable multivibrator timing period is determined by the unknown capacitor value. The monostable on period opens a gate so that pulses from a clock may be counted. The larger the capacitor, the longer the gate is open and the higher is the count. It is a simple matter to calibrate the count in terms of the monostable capacitor value
- a potential divider arrangement, as shown in Figure 12.13 where a standard capacitor C_1 is used with the unknown capacitor C_u. The output voltage is then indicative of the capacitor values, and the unknown capacitor may be calculated from the potential divider expression:

$$C_u = \frac{V_{in}\, C_1}{V_{out}} - C_1$$

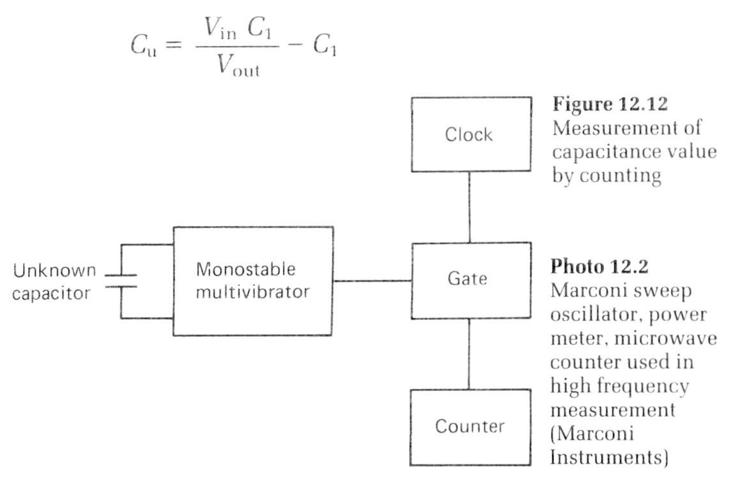

Figure 12.12 Measurement of capacitance value by counting

Photo 12.2 Marconi sweep oscillator, power meter, microwave counter used in high frequency measurement (Marconi Instruments)

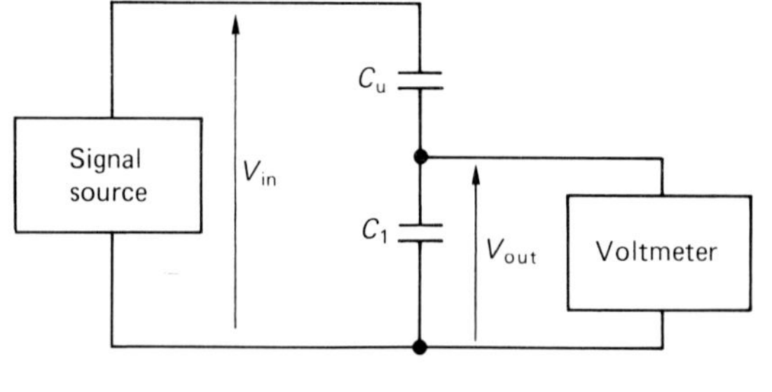

Figure 12.13 A standard capacitor, used with an unknown capacitor to form a potential divider, allows calculation of the unknown capacitor value

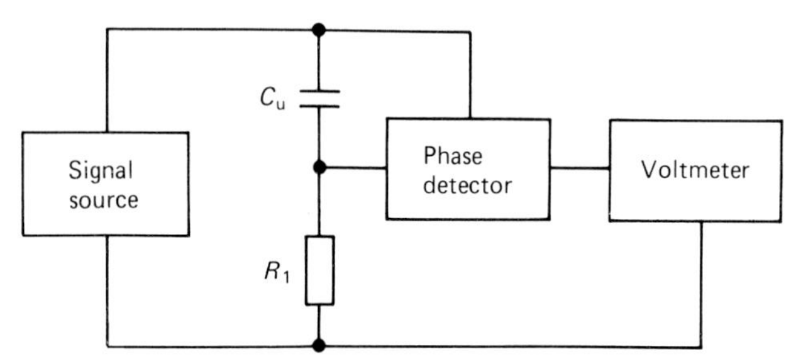

Figure 12.14 A potential divider, formed by a resistor with an unknown capacitor allows calculation of the unknown capacitor value

- a potential divider arrangement with a resistor, as shown in Figure 12.14. To avoid potential errors due to inductance of the resistor, a phase detector is used
- a Q meter, in the arrangement shown in Figure 12.15. Initially, the Q meter is setup with variable capacitor VC at a high capacitance, say, VC_h. The internal oscillator is adjusted to obtain resonance. The unknown capacitor C_u is connected in parallel with the variable capacitor, and the variable is adjusted to give resonance once again at a lower capacitance VC_l. The value of

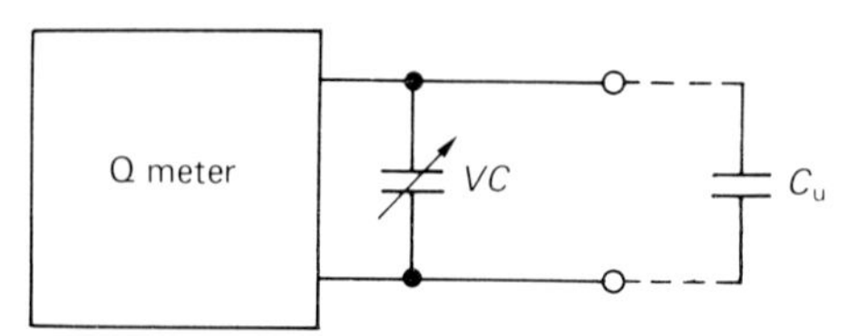

Figure 12.15 Using a Q meter to calculate an unknown capacitor's value

the unknown capacitor is obtained from the expression:

$$C_u = VC_h - VC_l$$

The Q meter also allows measurement of other capacitor parameters, as the Q factor may be measured when resonance is obtained at high (Q_h) and low (Q_l) values of the variable capacitor. For instance, the dissipation factor D of the unknown capacitor may be determined from the expression:

$$D = \frac{Q_h - Q_l}{Q_h Q_l} \times \frac{C_h}{C_h - C_l}$$

while the shunt resistance R_{sh} is obtained from the expression:

$$R_{sh} = \frac{Q_h Q_l}{Q_h - Q_l} \times \frac{1}{2\pi f_r C_h}$$

where f_r is the resonant frequency.

Although capacitance meters are available with the specific function of measuring capacitance, multimeters often incorporate one or more circuits to allow measurement of capacitance along with their other uses.

Current

Although an ammeter is usually used to directly measure current in a circuit, calculations of current (using Ohm's law) may be made indirectly using, say, an oscilloscope to measure the voltage created by a current through a known resistance. Higher currents (over 10 A, say) may be measured with clip-on devices which use magnetically sensitive techniques to determine the magnetic field setup by the flow of electric current.

Diode parameters

Many diode parameters are measurable. Some of the most common measurements are now described.

Breakdown voltage

A diode's breakdown voltage may be measured in many ways, including:

- with an ammeter and voltmeter, as shown in Figure 12.16. By forcing a constant reverse current at the value of the maximum permissible leakage current through the diode the voltage across it equals the breakdown voltage. This is a static measurement

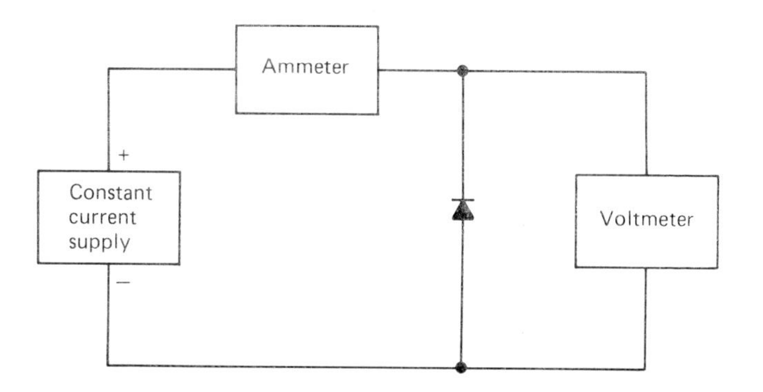

Figure 12.16 Static measurement of a diode's breakdown voltage

- with an oscilloscope, as shown in Figure 12.17. Here the voltage across the diode is applied to the horizontal input of the oscilloscope, while the current through it is monitored by applying the voltage obtained across a series resistor to the vertical input of the oscilloscope. This gives a dynamic measurement and display of the diode characteristics.

Reverse recovery time

A dual-trace oscilloscope is used in the arrangement of Figure 12.18 to measure diode reverse recovery

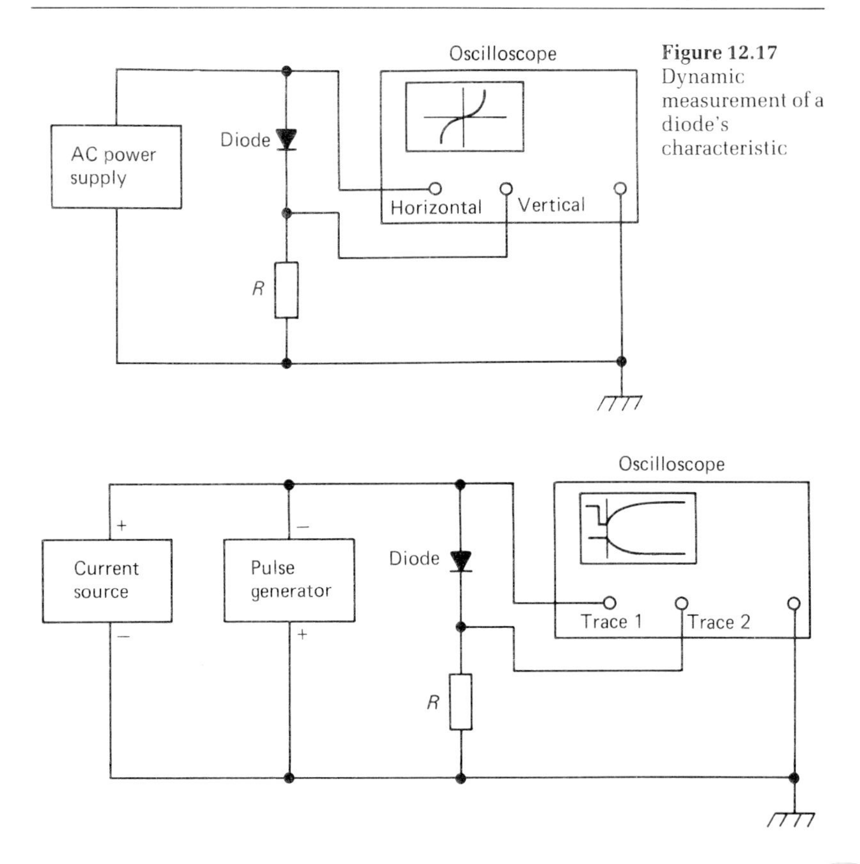

Figure 12.17 Dynamic measurement of a diode's characteristic

Figure 12.18 Measurement of a diode's reverse recovery time

time. One trace of the oscilloscope is used to display voltage across, while the other trace displays current through, the diode.

Voltage drop

Forward voltage drop across a conducting diode is simply measured:

- with an ammeter and voltmeter, in a similar arrangement to that in Figure 12.16, except the diode should be forward biased
- with an oscilloscope, as in the circuit of Figure 12.17.

Zener diode dynamic impedance

An AC ammeter and voltmeter may be used to measure a zener diode's dynamic impedance. Initially, the constant current is set to bias the zener diode to the required zener current. Then an AC current is superimposed across the zener via the coupling capacitor and the AC current and voltage are measured. Dynamic impedance is calculated from V_{AC}/I_{AC}.

An alternative arrangement uses the oscilloscope circuit of Figure 12.17 to display the diode's characteristics. The dynamic impedance is found by estimation of the slope of the characteristic curve at the zener current which is required.

Distortion

Distortion measurements are common in audio and radio receiver circuits. Typically, these are undertaken with distortion analysers of one form or another.

Frequency

Signal frequency may be measured directly using a frequency meter, or indirectly using an oscilloscope to display the signal then estimating its frequency.

Often estimation involves first estimating the waveform period, as measured on the oscilloscope graticule, then calculating frequency as the reciprocal. An alternative approach is to apply a signal with a known frequency (from an accurate signal source) to the oscilloscope horizontal input, and the unknown signal to the vertical input, as shown in Figure 12.19a. At certain ratios between the two frequencies Lissajous figures are displayed, examples of which are shown in Figure 12.19b to e. So, by adjusting the signal source frequency until a known Lissajous figure is displayed, the unknown frequency may be accurately determined.

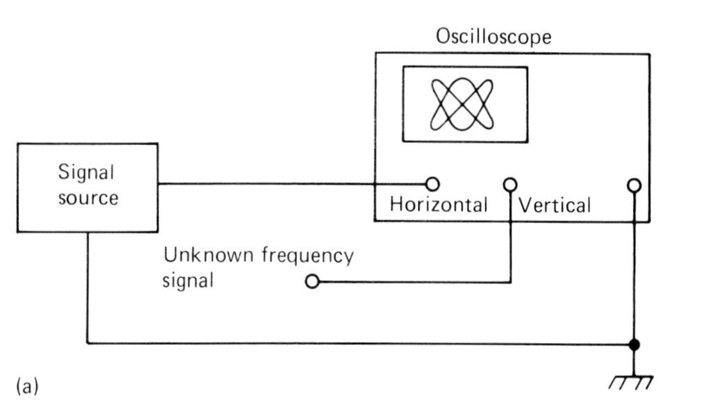

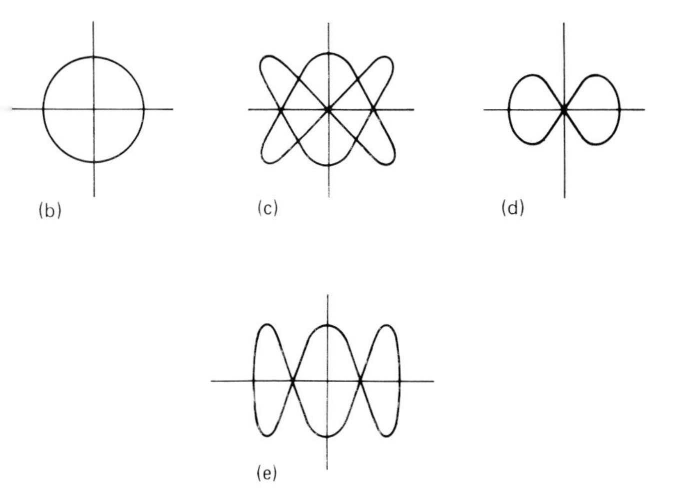

Figure 12.19 Calculation of an unknown signal frequency (a) using an oscilloscope to display Lissajous figures, at ratios of (b) 1:1 (c) 3:2 (d) 2:1 (e) 3:1

Lissajous figures are also frequently used to measure phase shift between signals.

Impedance

Like other basic measurands, impedance may be measured in a large number of ways. The most common are:

- with an ammeter, voltmeter and wattmeter, as shown in Figure 12.20. For a power P, current

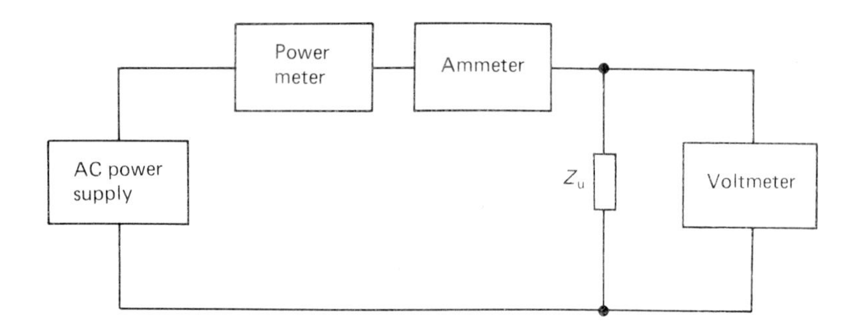

Figure 12.20 Measurement of impedance using power meter, ammeter and voltmeter

I and voltage V, the value of the unknown impedance Z_u is calculated from the expression:

$$Z_u = \frac{V}{I}$$

AC resistance in the impedance is calculated from the expression:

$$R = \frac{P}{I_2}$$

while the reactance of the impedance is:

$$X = (Z_u^2 - R^2)^{1/2}$$

- using a Q meter. Initially the Q meter is set to resonance when the unknown impedance Z_u is shorted (that is, with the impedance connections linked), and values of the variable capacitance VC, Q factor and resonant frequency f noted. Values of VC and Q are taken again with the short removed (that is, with the impedance connected) at the same resonant frequency. Reactance of the impedance is then calculated from the expression:

$$X_u = \frac{VC_1 - VC_2}{2\pi f VC_1 VC_2}$$

while resistance of the impedance is calculated from:

$$R_u = \frac{VC_1Q_1 - VC_2Q_2}{2\pi f VC_1 VC_2Q_1Q_2}$$

and the unknown Q factor is:

$$Q_u = \frac{(VC_1 - VC_2)Q_1Q_2}{VC_1Q_1 - VC_2Q_2}$$

If the impedance is capacitive, $VC_2 > VC_1$ and the unknown capacitance is calculated from the expression:

$$C_u = \frac{VC_1\ VC_2}{VC_2 - VC_1}$$

while, if the impedance is inductive $VC_1 > VC_2$, and the unknown inductance is calculated from:

$$L_u = \frac{VC_1 - VC_2}{(2\pi f)^2\ VC_1\ VC_2}$$

and if the impedance is resistive (obvious because no adjustment of the variable capacitor is required during measurement), the unknown resistance is:

$$R_u = \frac{Q_1 - Q_2}{2\pi f VC_1Q_1Q_2}$$

- using test equipment specifically made to ease impedance measurements, such as a vector impedance meter. Use of such equipment can be as simple as merely connecting the unknown impedance and turning on the meter – the equipment then displays all unknown values of the impedance.

Inductance

Measurement of inductance is usually undertaken using similar arrangements to those in the measurements of capacitance and impedance, usually as part of the same test equipment. Common methods include:

- using the substitution bridge used for measurement of unknown capacitance (Figure 12.11 on page 215). Procedure is identical to that of capacitance measurement, except replacing the unknown inductance for the capacitance and noting the balancing frequency. Inductance is then calculated from the expression:

$$L_u = \frac{R_2}{(2\pi f)^2 R_1 VC_a}(1 - VC_a/VC_b)$$

- using a phase detector method, similar to that for capacitance (see page 218), shown in Figure 12.21, where a resistor is placed in series with the unknown inductor to form a potential divider arrangement
- using a Q meter, described in the measurement of capacitance (see page 218).

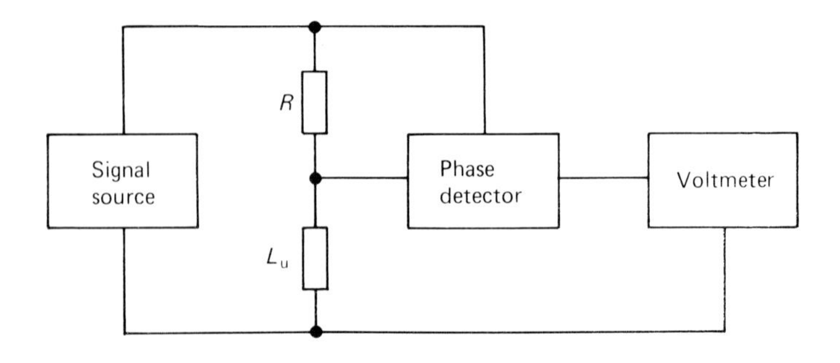

Figure 12.21 Inductance measurement using a phase detector and potential divider arrangement

Light
See *optical fibres*

Noise

Basic noise in an analog circuit is denoted in a number of ways. Consequently there are a number of methods used to measure noise, too. As most of the ways noise is denoted are related, however, it is often a job of taking only one or two measurements and calculating any others required.

First important noise parameter is a circuit's signal-to-noise ratio, which is given by the expression:

$$\frac{\text{signal power}}{\text{noise power}}$$

Power associated with the two signals can be found by measuring the signal and noise voltages, and calculating their mean square values then dividing each by the circuit's output resistance. Thus, the signal-to-noise ratio is:

$$\frac{\overline{V_s^2}/R}{\overline{V_n^2}/R} = \frac{\overline{V_s^2}}{\overline{V_n^2}}$$

where the line above the squared voltages indicates the mean value.

Because the signal-to-noise ratio is a power ratio it is usually expressed in decibels, where:

$$\text{signal-to-noise ratio (in dB)} = 10 \log_{10} \frac{V_s^2}{V_n^2}$$

The signal-to-noise ratio required by a system to perform well is very much dependent on the system. A hi-fi system should have a signal-to-noise ratio of at least 70 dB or more, so background noise is not heard between music tracks of an LP, say. That of a telephone is not so important; about 40 dB is adequate. An excellent television picture will be possible with an aerial signal-to-noise ratio of 50 dB. Nevertheless, knowing a system's signal-to-noise ratio allows comparison with similar systems.

Noise figure F, sometimes called *noise factor*, of a circuit is given by the ratio:

$$F = \frac{\text{input signal-to-noise ratio}}{\text{output signal-to-noise ratio}}$$

and because the noise figure, like signal-to-noise ratio itself, is a power ratio it is also commonly given in decibels where:

$$F(\text{dB}) = 10 \log_{10} \frac{\text{input signal-to-noise ratio}}{\text{output signal-to-noise ratio}}$$

However, as input and output signal-to-noise ratios are almost always expressed in decibels anyway, the noise figure (in decibels) may be calculated as:

$$F(\text{dB}) = \text{input signal-to-noise ratio (dB)} - \text{output signal-to-noise ratio (dB)}$$

The lower the noise figure, the better the noise performance of the circuit.

The noise figure of a system can be defined in terms of temperature, also. With temperature T in degrees kelvin, noise figure is given by the expression:

$$F = 10 \log_{10} \left(1 + \frac{T}{290}\right)$$

Where a system comprises a number of circuits or parts, overall noise figure is a combination of noise figures and power gains of each part (notation: F_1, F_2 and so on; P_{G1}, P_{G2} and so on), according to the expression:

$$F = F_1 = \frac{F_2 - 1}{P_{G1}} + \frac{F_3 - 1}{P_{G1} \cdot P_2} + \frac{F_4 - 1}{P_{G1} \cdot P_{G2} \cdot P_{G3}} + \ldots$$

This calculation can be extended to include any number of parts, as long as the power gains and noise figures of all parts are known. It's important to remember that *all* parts (amplifiers, attenuators, cables, transducers and so on) need to be included in the calculation, however.

It should be apparent from all of this, that most complex noise parameters can be expressed in terms of and derived from the simple parameter of voltage; the usual methods of voltage measurement should be referred to. Direct measurements of power and power gain (see page 213), and indeed noise figure itself will speed up these calculations, however. Noise figure meters exist to enable direct measurement of noise figure.

Optical fibres

Increasing use of optical fibres in communications systems has prompted a surge in the usage of associated test equipment. Most of the measurands which need to be measured, however, remain similar and analogous to those of non-optical fibre communications.

Attenuation

- fibre attenuation A is simply a matter of measuring the input power and the output power of the light using an optical power meter, and then calculating it from the expression:

$$A = 10 \log_{10} (P_{out}/P_{in})$$

Methods of doing this however require that power be measured in, often, impractical places in an optical fibre network. These are all frequency domain measurements

- one method, on the other hand, is a time domain measurement. Known as the *backscatter method*, it relies on the fact that light power injected into the fibre at one point is reflected back after an interval of time. The optical time domain reflectometer (OTDR) uses the backscatter method to create a visual display of injected pulse, reflections and time delays such that it is possible to determine attenuation as well as isolate imperfections such as poor connections in the fibre.

Bandwidth

Optical fibre bandwidth is found using two basic methods:

- in the frequency domain, with a spectrum analyser or power meter and signal source, simply sweeping a frequency through the range while measuring output power
- in the time domain, using an optical time domain reflectometer.

These are directly analogous to frequency response measurements in non-optical communications systems (such as untuned amplifiers – see page 211).

Optical time domain reflectometers are becoming increasingly more complex, and able to directly measure more and more measurands within optical fibre systems. Faults, or problems in the fibre, can be isolated down to the nearest millimetre or so using modern equipment, and it seems likely that OTDRs will be used almost exclusively for optical fibre measurements in the near future.

Phase

Measurement of phase or at least phase difference between two signals is often undertaken on the oscilloscope, in two main ways:

- displaying both signals on the oscilloscope screen, as in Figure 12.22, allowing measurement of wave period T, and time difference between the two signals t. Phase difference P may then be calculated in degrees according to the expression:

$$P = \frac{t}{T} 360$$

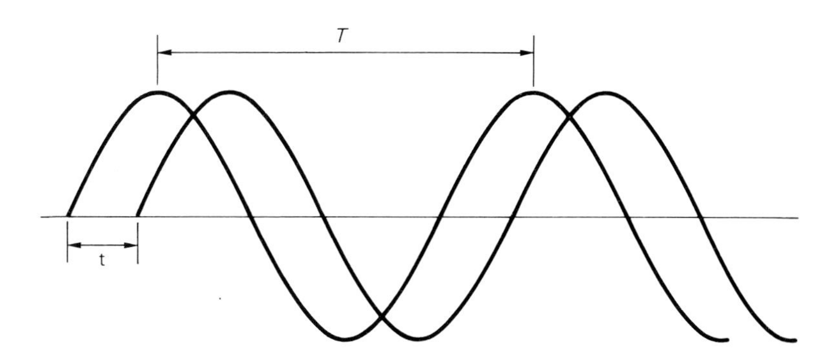

Figure 12.22 Using an oscilloscope display of two signals to estimate phase shift between them

- displaying a Lissajous figure, applying the reference signal to the horizontal input of the oscilloscope and the unknown signal to the vertical input. Vertical and horizontal amplifiers should be adjusted so that the figure fits into a square. The phase shift between the two signals will be apparent on the display as the angle to which the figure tilts, as shown in Figure 12.23a. The phase shift is calculated as the ratio of the figure

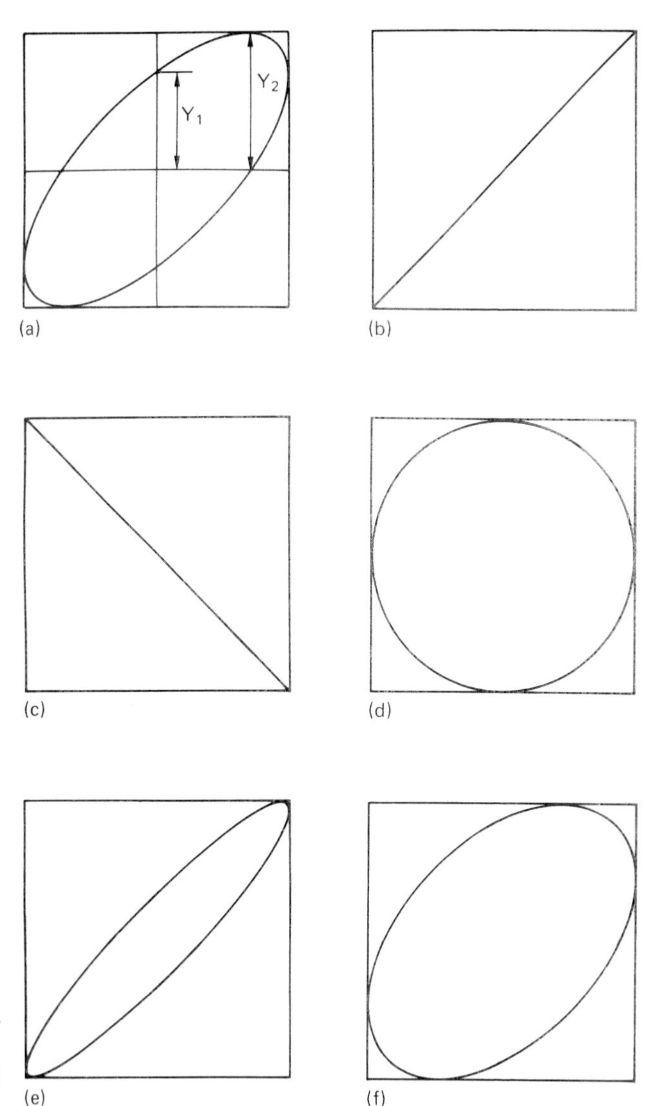

Figure 12.23 Lissajous figures of phase shifted signals (a) measurements to be used in calculation (b) phase shift of 0° (c) phase shift of 180° (d) phase shift of 90° or 270° (e) phase shift of 20° (f) phase shift of 60°

dimensions, such that the phase shift θ is given by the expression:

$$\theta = \sin^{-1} \frac{Y_1}{Y_2}$$

A selection of displayed Lissajous figures, with a variety of phase shifts, is shown in Figure 12.23b to f.

Pulse parameters

A number of pulse parameters may need to be measured. Details of the most common follow.

Duty cycle, pulse width

A basic display of a pulse waveform on an oscilloscope will give a rough estimation of pulse width and duty cycle – to at best about ± 5%. A possible display is shown in Figure 12.24, where the pulse period T is shown, along with its pulse width t. Pulse duty cycle D as a percentage is then calculated from the expression:

$$D = \frac{t}{T} \cdot 100$$

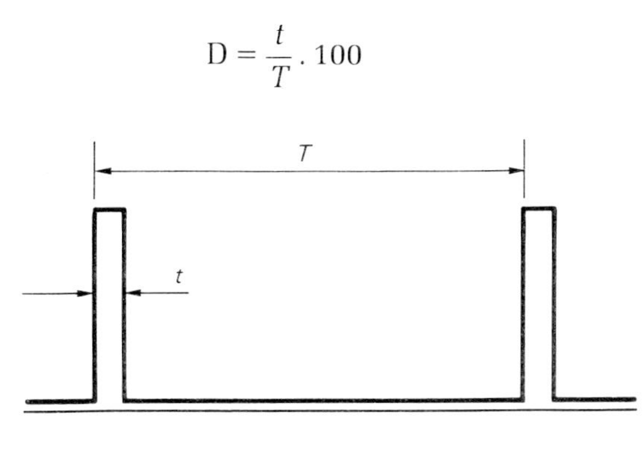

Figure 12.24 Estimation of pulse width and hence duty cycle on an oscilloscope display of pulse waveform

Estimation using this method may not be accurate enough, particularly if the duty cycle is very small. A more accurate measurement of duty cycle (or any time on an oscilloscope display) may be made using the *delay time* control of advanced oscilloscopes. With one trace intensified onto the other, the delay time control allows the intensified part to be measured more accurately. For small times, too, the inherent rise times of the measurement equipment must be taken into account (see the section on measurement of time using an oscilloscope (page 236).

Fall time, rise time
Either of the methods used for estimation of duty cycle may be used for estimation of rise and fall times of a pulse, calculated as the time taken for the pulse to rise (or fall) between 10% and 90% (or 90% and 10%) of the amplitude.

Pulse rate
Pulse rate can be measured in a number of ways, including:

- using an oscilloscope to measure the pulse waveform period, calculating the reciprocal of this time to obtain the frequency
- using a frequency time and event counter to measure the frequency directly.

Resistance
One of the basic electrical measurands which require measurement, resistance is usually measured by connecting a known voltage across the unknown resistance and measuring the current through it. From Ohm's law: $R = V/I$, the resistance can then be calculated. Generally speaking, most equipments to measure resistance somehow perform this calculation internally, so users need only to read the resistance value.

A problem associated with measurement of resistance is that rarely is a 'resistor' only purely resistive. Normally, it is an impedance which, for the main part, *is* resistive but includes capacitive and inductive parts. Under DC conditions these parts play no effect, so true resistive measurement is possible. Under AC conditions, however, the capacitive and inductive parts are relevant – particularly at high frequencies.

Common methods used to measure resistance include:

- the ohmmeter, which is usually included as part of a multimeter. The multimeter has an internal voltage source which creates a current through the unknown resistance, the value of the current being inversely proportional to the resistance (from Ohm's law). An ammeter arrangement within the multimeter is used to measure the current
- a bridge circuit, such as the substitution bridge described on page 215.
- a phase detector, similar to those used in measurement of capacitance (page 218) and inductance (page 226), as shown in Figure 12.25. An AC signal is applied across the unknown resistance R_u and a reference resistance R_r known to have negligible impedance. The resultant voltages (V_u across R_u, and V_r across R_r) are compared by the phase detector, the output of which comprises only those parts of V_u which are in phase with V_r (Figure 12.25). The output voltage is therefore a true indication of the applied voltage across the resistor

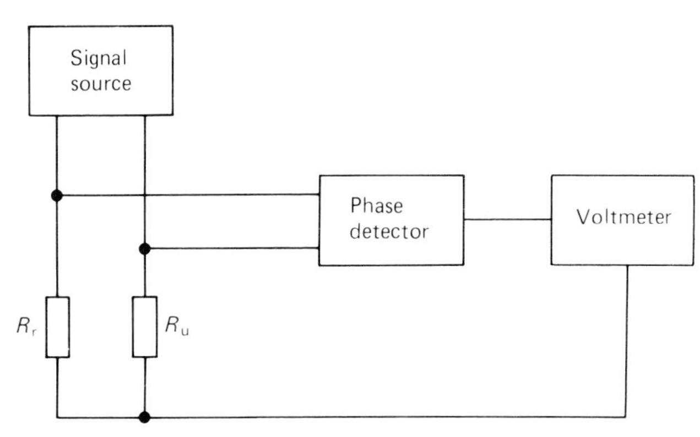

Figure 12.25 Using a phase detector to accurately compare voltages across unknown and known resistances

- a Q meter, used in a similar way to its measurement of impedance (page 224). This method is useful for measurement of capacitive and inductive parts of the resistance at high frequency.

SCR parameters

Most important parameters in SCR devices such as thyristors and triacs are successfully measured using simple equipment. Figure 12.26 shows a basic arrangement which may be used to measure thyristor anode and gate characteristics statically. The arrangement is adaptable to suit and, by adjusting voltages at the gate and anode, most characteristics such as anode forward breakdown voltage, leakage current, holding current, latching current, and gate turn-on voltage and current can be measured. By observing the gate and anode voltages on an oscilloscope screen, turn-on time can be estimated, too (see below for details of using an oscilloscope to measure time).

Signal-to-noise ratio

See the section on *noise* (page 227) for details abouts signal-to-noise ratio and measurement of noise itself.

Time

Frequency, time and event counter

Time interval between, say, two pulses is most conveniently measured using a frequency, time and event counter using the direct gated counter principle (see page 82).

Oscilloscope

An alternative is to use an oscilloscope (see pulse parameters, page 233) to estimate the period directly off the display. Simple estimation like this is acceptable for relatively long measurement periods, but is not particularly accurate (no better than about $\pm$ 5%).

Smaller times, say, rise or fall times of individual pulses which need to be accurately measured, are best estimated on the oscilloscope using the *intensify*

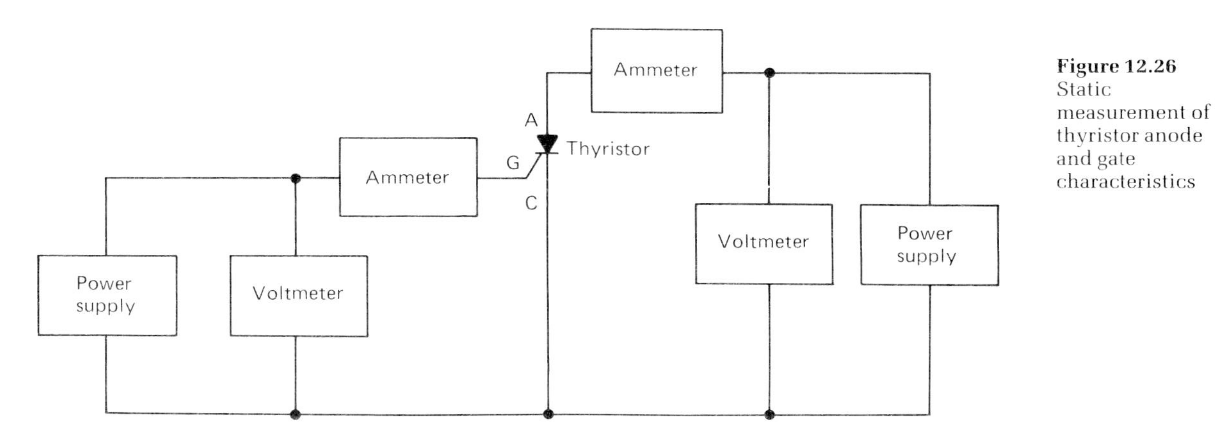

Figure 12.26 Static measurement of thyristor anode and gate characteristics

Photo 12.3 Tektronix digitising oscilloscope in use (Tektronix)

and *delay time* facilities. By intensifying a selected part of one trace onto the other using the delay time control (which, incidentally, features a 10-turn calibrated potentiometer) the selected part of the signal can be effectively magnified so that pulse edges can be viewed in comparitive detail. This allows a much more accurate estimation of small time periods.

Rise (and fall) times of measurement equipment such as oscilloscopes and signal sources should be borne in mind when attempting to measure extremely small time periods, say rise time introduced by a circuit as shown in the arrangement of Figure 12.27. Here, although the circuit being tested has a distinct rise time as measured on the oscilloscope, it may not be as big as the oscilloscope display would suggest. If, say, the signal source rise time is t_s, the oscillo-

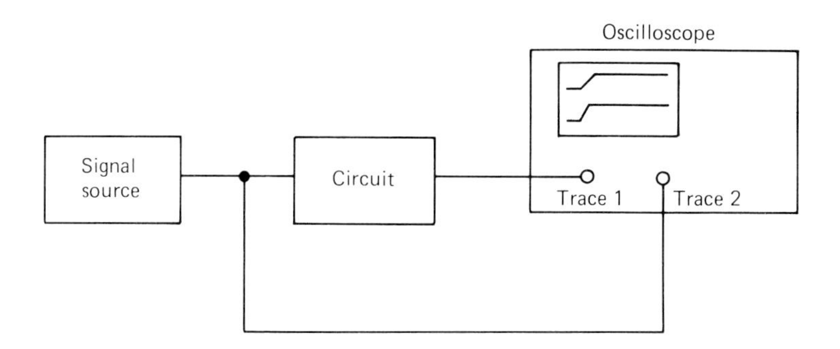

Figure 12.27 Using an oscilloscope and signal source to accurately measure rise time introduced by a circuit – allowing for rise times of test equipment

scope rise time is t_0, and the rise time measured on the oscilloscope display which is apparently introduced by the circuit is t_c, then the *actual* rise time t_a introduced by the circuit is given by the expression:

$$t_a = (t_c^2 - t_o^2 - t_s^2)^{1/2}$$

Introducing some values into this expression, say, $t_c = 30$ ns, $t_0 = 10$ ns, $t_s = 10$ ns, then the actual rise time introduced by the circuit is in the region of 26 ns – an error in the apparent measurement of around 13%.

Transmission lines

There are two main characteristics of transmission lines which need to be regularly measured.

Cable fault location

Two methods of fault location in cable systems are commonly used, and can give considerable accuracy:

- a bridge measurement, illustrated in Figure 12.28, where a cable of length l is shown with a fault at a distance of l_f from the measurement end. Fault is a low resistance connection to ground – which is the most common form of fault on such cable networks. To perform the test, the other end of the

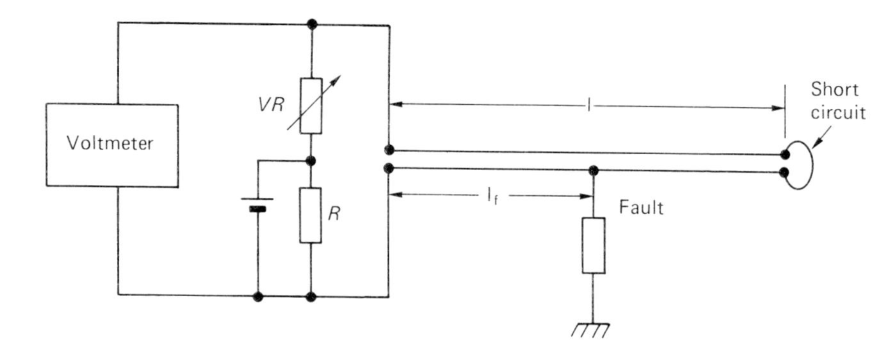

Figure 12.28 Bridge measurement of cable fault location

cable is short-circuited, such that a bridge circuit is obtained which uses the resistance of the cable itself as two arms of the bridge. The fact that conductors are assumed to have uniform cross-section over their lengths means that calculations of their resistances are directly proportional to cable lengths. Given this assumption, once the bridge is balanced, the fault will be located upon calculation of the expression:

$$l_f = \frac{2l.VR}{VR + R}$$

Other bridge methods exist, too, capable of considerable accuracy (to around ± 0.5%)

- time domain reflectometry. Any signal entering the cable takes a certain time to travel away from the equipment, be reflected from the cable fault, then travel back. A basic time domain display of this on a time domain reflectometer records the time as a distance along the horizontal axis – the distance along the axis being proportional to the time, which in turn is proportional to the cable distance. In effect, the display is a direct record of cable length. Further, the appearance of the signal received gives information about the way it was reflected back and the fault itself. Basic time domain reflectometer displays are illustrated in Figure 12.29a to d.

Standing wave ratio
An ideal match between a transmitter and an aerial occurs when the characteristic impedance of the transmission line cable Z_C equals the characteristic impedance of the aerial load Z_l. In such a case, the forward wave from the transmitter is totally absorbed by the aerial and no wave is reflected. There is then said to be no standing wave. The *standing wave ratio* (SWR) is a measure of the transmission system; it is the ratio of the characteristic cable impedance to the characteristic load impedance, such that:

$$\mathrm{SWR} = \frac{Z_C}{Z_l}$$

so, for an ideal system, the SWR is 1:1.

As these impedances act directly on the voltages and powers of the forward and reflected waves in the transmission line (V_f, V_r, P_f, P_r), it is possible to equate them with the standing wave ratio, too, such that:

$$\mathrm{SWR} = \frac{1 + V_r/V_f}{1 - V_r/V_f}$$

and;

$$\mathrm{SWR} = \frac{1 + (P_r/P_f)^{1/2}}{1 - (P_r/-_f)^{1/2}}$$

Consequently, many methods can be used to measure SWR, including bridge circuits, power measurements, spectrum analysers. Many types of direct reading SWR meters are available also.

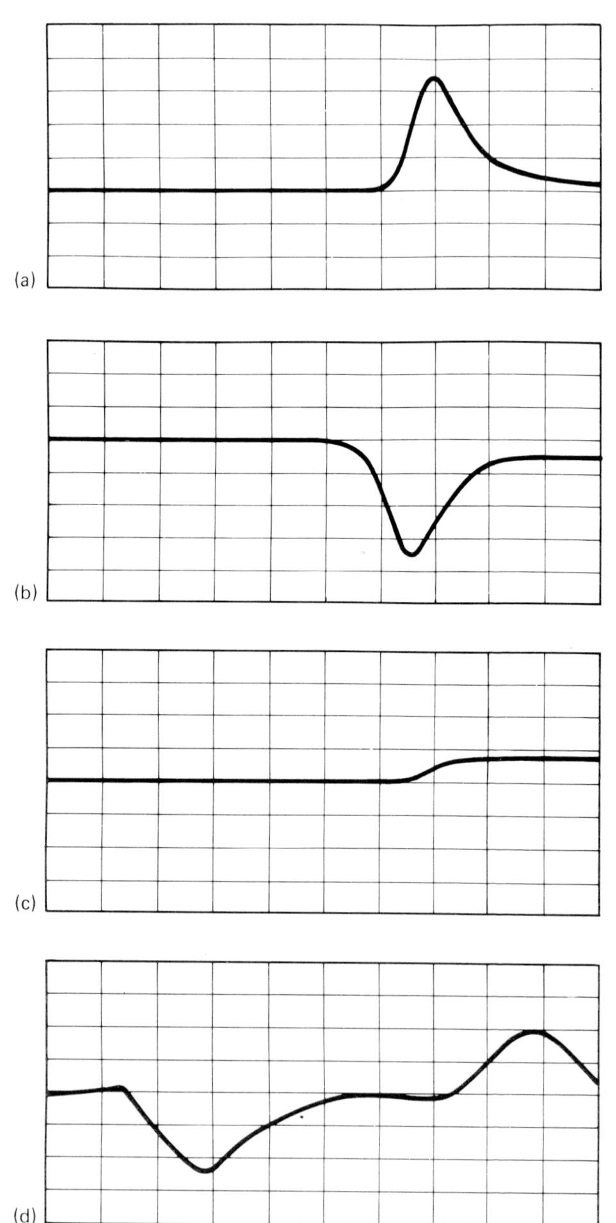

Figure 12.29 Basic time domain reflectometer displays of cable faults (a) inductive or higher resistance faults (b) capacitive or lower resistance faults (c) a joint where higher impedance cable is added (d) water in the cable

Appendix 1
Displays

The purpose of displays in electronic test and measuring equipment is to inform the user of the measured conditions of the tested circuit. It's reasonable to think, therefore, that the type of display used depends very much on the test undertaken. And, because there must be so many different kinds of tests and so many different kinds of test instruments, there should correspondingly be a large number of types of display. Fortunately, this is not the case, as many different tests can use the same, or similar, type of display. Most displays are of three categories: analog (the moving-coil movement type); digital (liquid crystal, light-emitting diode, etc.); or a combination of analog and digital (cathode ray tube). Identifying these three broad categories of displays makes it easier to look at and compare the available types.

Analog displays – the moving-coil movement

The moving-coil meter movement is the veteran of display devices. In one form or another it has been around for over 150 years, although it was only about 100 years ago that the first moving-coil movements with all the attributes common to modern movements was made.

Moving-coil movement operation depends on the fact that when a current passes along a wire lying in a steady magnetic field, the magnetic field which the current itself creates interacts with the steady magnetic field, so a force occurs tending to move the wire. Movement of the wire is perpendicular to the directions of current flow and steady magnetic field.

Early attempts at a meter movement relying on the phenomenon mostly used mirrors to indicate position by light reflection, but most movements manufactured today use an indicating pointer. Construction of a moving-coil meter movement is illustrated in Figure A1.1, but it must be emphasised that this is only a simplified construction to show the principle; many variations occur in actual moving-coil movements.

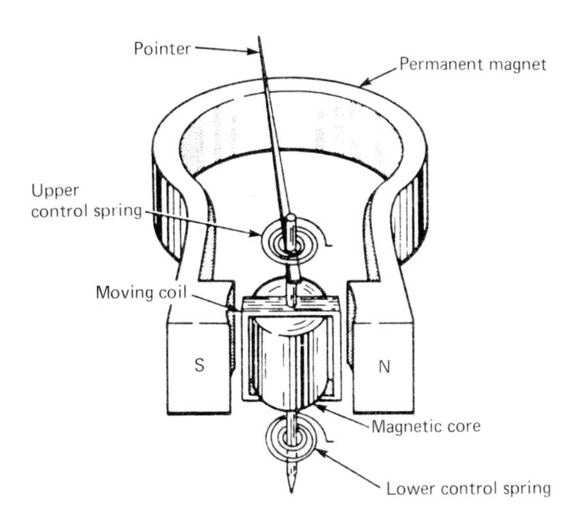

Figure A1.1 Basic construction of a moving-coil meter movement

The wire carrying the current to be measured is wound into a rectangular coil around a core of magnetic material such as soft iron. The core is pivoted so that it can rotate, and mounted with springs, so that it returns to a defined position. A pointer, attached to the core and mountings, is used to indicate the rotational position of the core. In a complete meter, an indicating scale is positioned underneath the pointer so that pointer position can be read.

The core and coil arrangement is mounted in the magnetic field produced by a permanent magnet. Magnet shape, together with the magnetic core in the coil, ensures that, whatever position the core and coil rotates to, the permanent magnetic field is parallel to the coil plane, therefore the strength of the magnetic field remains constant. When direct current passes through the coil it generates another magnetic field, the strength of which is dependent on the size of the current, which directly interacts with the permanent magnetic field. Reaction of the two fields establishes a torque, which rotates the coil and core until the torque produced by the now twisted springs is the same size. The amount rotated therefore depends on the size of the current through the coil, so pointer position on the scale indicates the current value. Note that application of an alternating current through the coil produces an alternating torque, tending to rotate the coil and core first in one direction then in another. Inertia of the movement prevents the coil and core from actually moving for all but very slowly alternating currents, however, so no rotation occurs. In other words, the moving-coil meter movement does not indicate alternating current as it stands here, and must be adapted to do so.

So far, we have only considered what happens to the movement when current passes through the coil. As the coil has resistance, however, the same effect will be observed if a voltage is applied across the meter. So the movement can be used to indicate values of voltage as well as current.

The moving-coil movement has a number of advantages, but also some severe limitations, which define its use in test equipment. A significant advantage is that it derives its driving current or voltage from the circuit under test. No power supply with its mains source requirements is needed. A simple multimeter comprising the moving-coil movement and associated circuitry is a versatile piece of test equipment; totally portable and usable in almost any ambient condition.

On the other hand, the movement's low resistance may significantly load the circuit under test, producing incorrect readings. Accuracy is not particularly good and depends, partially at least, on the user's ability.

Analog/digital displays – the cathode ray tube (CRT)
Figure A1.2 shows a simplified cathode ray tube made up from an evacuated glass tube, widened and flattened at one end to form a viewing screen. The inside of the screen is coated with phosphor, a material which is fluorescent and glows when electrically excited, say when hit by electrons. A number of electrodes are positioned inside the tube to perform various functions. The main electrodes and their functions are:

- a heater – literally a coil of wire which is heated to glowing point by the application of an electric current at low voltage. The area around the heater becomes rich in electrons
- an acceleration anode electrode – held at a high positive potential (with respect to the heater) so that it attracts electrons from the heater. As the electrons reach this electrode their kinetic energy is such that they pass right through the electrode

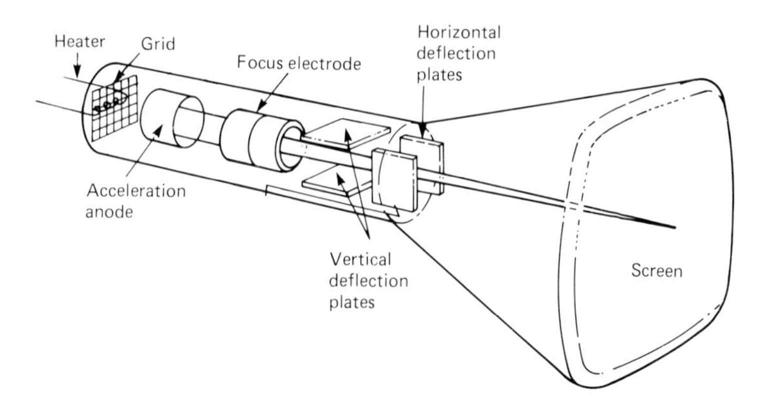

Figure A1.2 Basic construction of a cathode ray tube

and towards the front of the tube, in a beam (the 'cathode ray')

- a control grid – which is held at a variable negative potential to the heater, thus repelling or not repelling the negative electrons and allowing a controlled amount through
- a focusing electrode – the potential on this electrode is variable so as to create a beam with as small a cross-sectional area as possible when it strikes the screen. The heater, accelerating electrode, control grid and focusing electrode are often collectively known as an electron gun
- deflection plates – the potential across each pair of plates (vertical and horizontal), and the polarity of the potential (that is, positive or negative) defines how much the electron beam is deflected away from the centre of the screen. If, say, the upper plate is positive and the lower plate is negative, the beam will be deflected up (towards the positive plate and away from the negative plate). Similarly, by varying the potentials applied to the horizontal deflection plates the beam is deflected in a left or right direction.

So, the beam direction (and thus the position where it hits the screen), strength (that is, the brightness of the trace), and width (the thickness of the trace) are all controllable electronically, merely by changing the voltages applied to the various electrodes. Many refinements may be added to this basic cathode ray tube to improve performance, sensitivity, or beam brightness, but the operation remains more or less as discussed.

Storage CRT

One refinement which almost changes the very nature of the cathode ray tube, on the other hand, is illustrated in Figure A1.3, where a storage-type

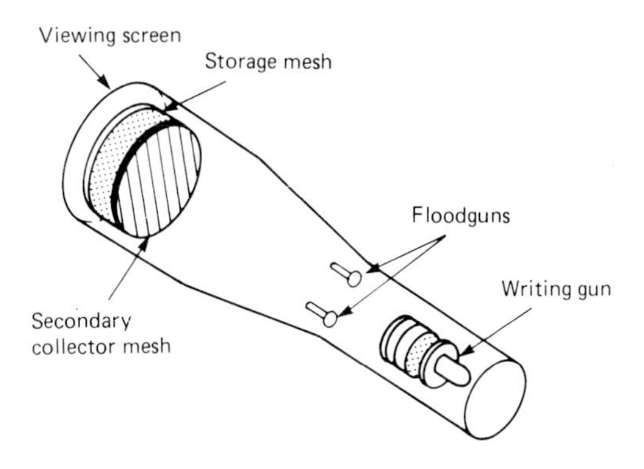

Figure A1.3 A storage cathode ray tube, which uses floodguns and a storage mesh to form a bistable-type display of a written trace

cathode ray tube is shown. An ordinary cathode ray tube, as described above, can only be used to display real-time occurrences; that is, measurands that take place continually, as the display is viewed. The storage cathode ray tube, however, is capable of displaying a single occurrence, long after it has been measured.

Just behind the phosphor screen is an assembly, known as the **target screen**, containing a fine **storage mesh**. Unlike a conventional tube, the storage cathode ray tube has two types of electron gun; the usual electron beam gun, and two or three other electron guns. The usual gun is known as the **writing gun**, and performs the accepted functions: a storage cathode ray tube can, in fact, be used as a real-time cathode ray tube. The **flood guns** are used to 'floodlight' the storage mesh with a broad parallel beam of electrons with a low velocity, charging the mesh to a uniform low negative potential. In this state, no electrons from the flood guns hit the phosphor screen.

On its way through the storage mesh to the phosphor screen, the writing gun beam causes secondary emission to occur from the mesh. More electrons are lost from the mesh than are gained from the writing gun beam, so the mesh becomes positively charged at the localised points which the writing gun beam hits. These local positively charged points attract the flood beam gun electrons toward them with a greater velocity, so that the flood beam electrons themselves cause secondary emission and pass through to the phosphor screen. In this way, a trace written once on the screen is 'stored', although to be precise no storage actually takes place, only a regeneration process which is triggered off by the original single trace.

The storage process is a bistable one, maintained indefinitely until the flood gun beam is reduced below the level where regeneration occurs, accomplished by altering a controlling voltage to the flood gun control grids. The storage cathode ray tube is capable of displaying a single measurand event, for as long as the user wishes to maintain it on the screen. Displayed events cannot be adjusted or repositioned in any way, however, once stored.

In general, the cathode ray tube is a most useful display device. It is capable of extremely fine resolution of detail, although accuracy is rather limited by the user's ability to read the displayed information, which is generally in analog form. However, some of the more recent cathode ray tube developments in test instruments such as oscilloscopes, logic analysers, spectrum analysers etc., may present the information in alphanumeric form, thereby eliminating reader error. A display may be read in most lighting conditions, as the cathode ray tube is an emissive display device; that is, it generates its own light. Main disadvantages are size, weight, and power requirements.

Digital displays

There are a number of types of digital display. The most common on recent test equipment are liquid crystal displays (LCD), light-emitting diode (LED) displays, and vacuum fluorescent displays (VFD). Their similarity lies in how they actually display the information – normally in a seven-segment or alphanumeric manner – rather than how they work. For this reason it's pointless to discuss operational characteristics of all the types, and only one digital display is looked at closely: the liquid crystal display.

The main principle that all liquid crystal displays follow is illustrated in Figure A1.4, where a layer of liquid crystals is shown sandwiched between two transparent electrodes. The molecules of the liquid crystals are generally aligned in the vertical direction and so light from behind the device can pass through. However, when an energising potential is applied across the electrodes, as in Figure A1.4b, the liquid crystal molecules are polarised into

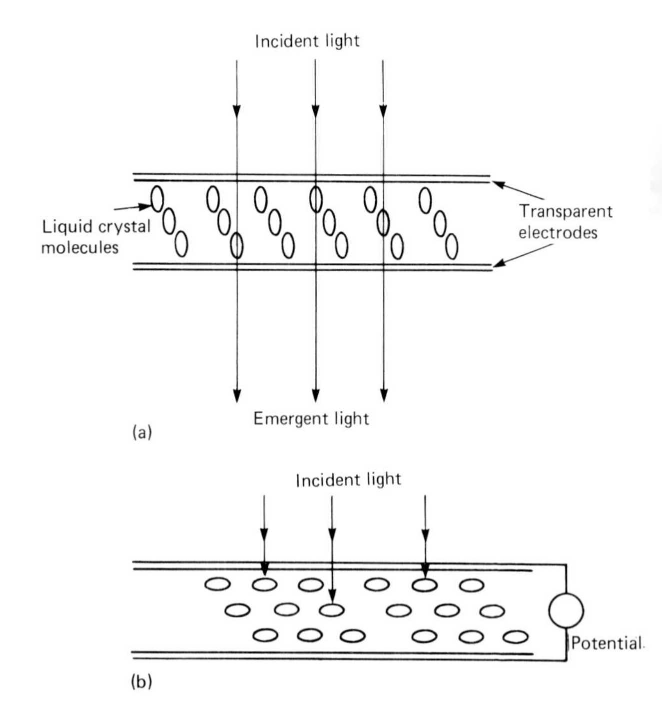

Figure A1.4 Principle of liquid crystal displays. A layer of liquid crystals is sandwiched between two transparent electrodes (a) when no potential is applied across the electrodes the molecules are polarised such that light can pass through the structure (b) when a potential is applied the molecules are orientated so that light cannot pass through

horizontal alignment, which prevents light from passing through. Generally, a layer of reflecting material is placed directly behind the rear electrode, so that incident light passes through and is reflected back in the unenergised state (giving the appearance of a white object), or does not pass through and so cannot be reflected back in the energised state (giving the appearance of a dark object). Such a display is non-emissive, and depends for its visibility on an adequate amount of ambient lighting. Sometimes, however, liquid crystal displays are used which have some form of back-lighting (that is, a light source positioned to the rear of the display), creating an emissive display.

Three main varieties of liquid crystal – nematic, cholestric and smectic – differing mainly in molecular alignment as shown in Figure A1.5, are used to make liquid crystal displays.

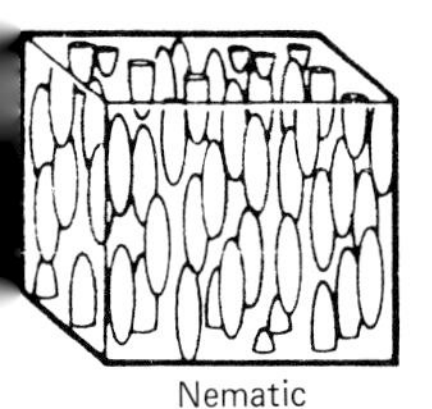

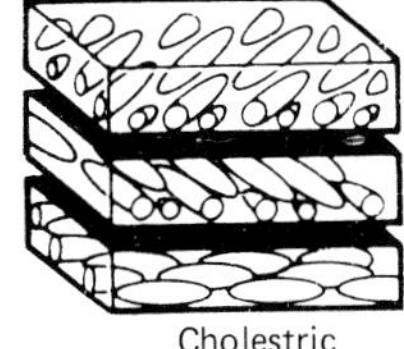

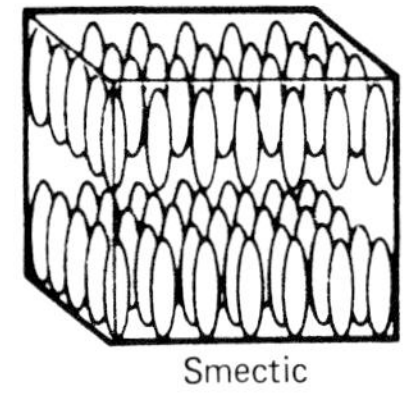

Figure A1.5 Three types of molecular structure in liquid crystal materials
(a) nematic
(b) cholestric
(c) smectic

Nematic liquid crystals are more commonly called 'twisted nematic' crystals, because the crystals are twisted through 90° between one electrode and the other when unenergised. Polarising sheets (at 90° to each other) are applied at the front and back of the device, so that incident light entering the display is polarised into one plane by the first polarising sheet, passes through the liquid where it is twisted through 90° by the structure, then leaves the device through the second polarising sheet. When an energising potential is applied, the molecules are all aligned in the same direction as the light, which now passes through the liquid crystal without being twisted, so cannot pass through the rear polarising sheet.

A cholestric liquid crystal display has no polarising sheets but a dichroic dye is added to the liquid crystal. Dye cells align themselves with the liquid crystal molecules, in effect producing an electronically controlled colour filter which changes colour with the applied voltage. This dye addition is the reason why displays using this principle are sometimes known as 'guest-host' devices; the dye cells are the 'guests' in the crystal 'host'.

Smectic liquid crystal displays rely on the different properties (that is, phases) of liquid crystals at different temperatures. In the smectic temperature phase, molecular arrangement of the liquid crystal cannot be changed by application of an energising potential. However, when heated to the nematic phase, then cooled to the smectic phase, orientation of the molecules will be as it was in the nematic phase. To exploit these phenomena, smectic displays use a combination of matrices of electrical lines, so that each individual display element can be addressed and *heated*, from their normal smectic state to the nematic phase. Operation of a smectic display is therefore something like that of a twisted nematic display, but it has a memory facility which retains and displays the last information input during the nematic phase.

Shape of the liquid crystal display depends purely on the display requirements, as the electrodes can be manufactured in almost any shape and size the user may require. For basic numerical applications (say a simple digital multimeter, which is to display a number of digits) a seven-segment arrangement is sufficient, but more complex alphanumeric or graphic information (to be displayed on a computer or an oscilloscope) requires a correspondingly more complex liquid crystal display, typically of dot-matrix form.

Advantages of liquid crystal displays, and indeed of all types of digital displays, in test equipment instruments are manifold. Most important, in a display of seven-segment or alphanumeric nature, accuracy and resolution depends not on the device, but on the measuring circuit that feeds the digital information to the display. Users' reading errors, inherent in the use of moving-coil or cathode ray tube analog displays, are eliminated for the same reason. Test instruments using digital displays are therefore typically much more accurate than analog equipment of the same price.

On the other hand, numerical digital displays are updated at rates of between about two or three times a second, allowing sufficient time for the user to read one measurement before the next is displayed. So, measurement of, say, a slowly varying measurand is not possible, and can only effectively be done with an analog display.

Liquid crystal displays, unlike most other forms of digital displays, have a very low power consumption, so can be built into portable battery-powered test equipment. In most applications, therefore, liquid crystal displays are eminently qualified.

Some applications, on the other hand, are not suited to liquid crystal displays. Being non-emissive they rely on a suitable level of background light, something which is of no consequence in the laboratory, but may be of significance elsewhere. Back-lit displays are however available, but this defeats the low power consumption argument. The other types of digital displays such as light-emitting diode or vacuum fluorescent will probably be more useful here, as they *are* emissive and can be easily viewed in the darkest of environments – but they have a consequently greater power consumption.

Appendix 2
Optical fibres

Optical fibres form a relatively new type of communications medium. Popularity, however, is growing extremely fast as they have a number of significant advantages over metal cabled systems, not the least the fact that a single fibre can be used to carry many times the amount of information which can be carried by a metallic cable.

The principle behind optical fibre communications is that fine strands of transparent material – generally glass – may be used to *guide* light between source and sensor. In this respect, an optical fibre strand acts as a type of waveguide.

A single optical fibre comprises an inner core of transparent material covered by a cladding layer, also transparent (Figure A2.1). Over this may be a jacket of polyurethane or similar non-transparent material. Total diameter of this fibre may only be about 0.1 mm or so.

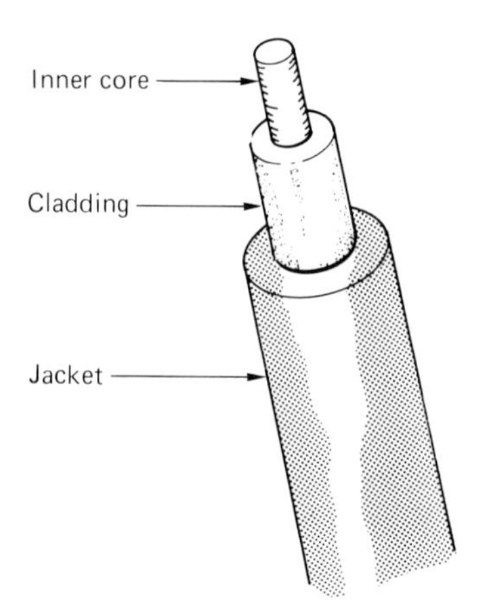

Figure A2.1 Construction of a single optical fibre

Generally, single fibres are not used, instead many are combined together into a cable. Many different types of these optical fibre cables are available, varying in cost and chosen according to application. There are also a number of different types of fibre, each with its own transmission characteristics, but all operating to the same principle.

Transmission of light

There are two important methods by which light is transmitted: reflection (in which light is *bounced* off a smooth surface) and refraction (in which light passes through a material). To understand how optical fibres work, these two methods must be considered. The first point, however, is to note that light may be reflected off glass, or refracted through it. The importance of optical fibres lies in the relationship between the reflection of light off glass and refraction of light through it.

In Figure A2.2a a ray of incident light is reflected off a glass surface. The light ray is incident at an angle θ to the perpendicular (known as the **normal**). The angle between normal and incident ray is known as the **angle of incidence**. The reflected light ray leaves the surface at an angle φ to the normal, the **angle of reflection**. The angles of incidence and reflection are equal.

In Figure A2.2b, on the other hand, light is incident onto a glass surface such that refraction occurs. Now, the **angle of refraction** φ is not equal to the angle of incidence θ.

This difference in angles is due to the fact that light travels faster in some materials than others. In the refraction example of Figure A2.2b, light travels first through air at one speed then through glass at another. In the reflection example of Figure A2.2a, however, light travels only in air – before *and* after reflection, so the angles are equal.

Figure A2.2 Incident light may be (a) reflected off or (b) refracted through glass

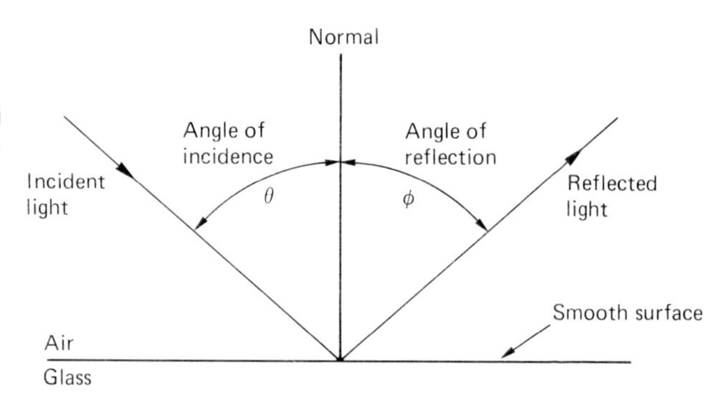

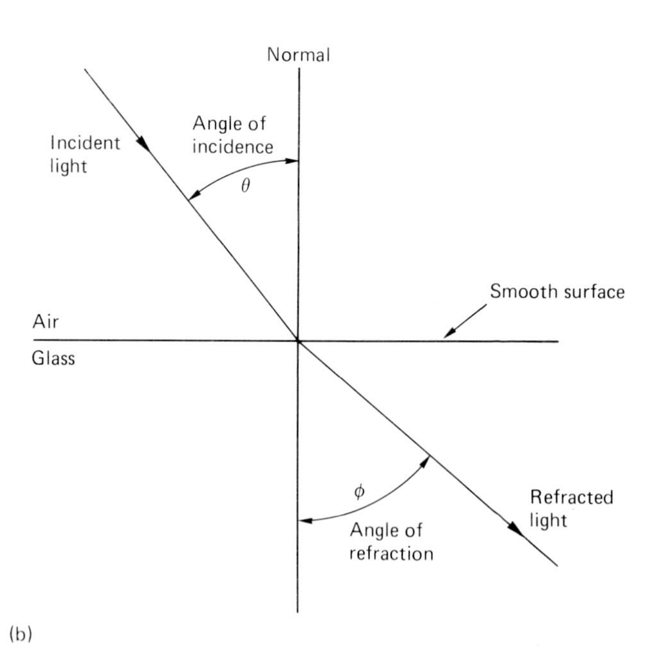

Ratio of the speed of light in a vacuum to its speed in any given material is known as that material's *index of refraction*. As light travels faster through a vacuum than through any other material, the index of refraction n will always be greater than 1, although for air it is so close to 1 that for most purposes it can be taken as unity.

The relationship between angles of incidence and refraction is expressed by *Snell's law of refraction*:

$$n_1 \sin \theta = n_2 \sin \varphi$$

where n_1 and n_2 are the indices of refraction of the materials light is passing through.

Given this relationship it is a fairly simple matter to determine the angle at which a glass surface no longer reflects light, instead refracting it. When light is refracted, the angle of refraction must be 90° or less (otherwise the light is simply *not* refracted but reflected). Figure A2.3 shows the changeover angle from reflection to refraction, which occurs when the angle of refraction $\varphi = 90°$. If the refractive indices of glass and air are 1.4 and 1, the angle of incidence can be calculated from Snell's law as:

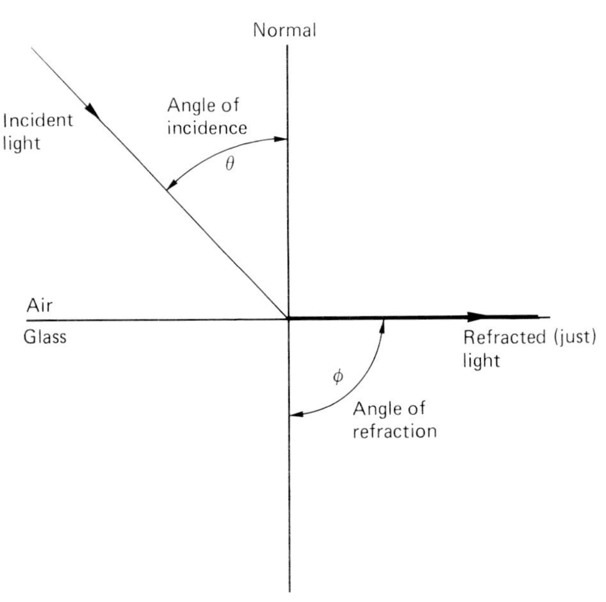

Figure A2.3 Changeover angle from reflection to refraction, where the incident angle is the critical angle

$$\sin\theta = \frac{1 \sin 90^\circ}{1.4}$$

So, $\theta = 45.58^\circ$. This is known as the **critical angle**; all light incident on the surface at an angle greater than this is reflected, light incident at a lower angle is refracted.

Figure A2.4 illustrates how knowledge of this critical angle is used in optical fibres. Here an LED light source transmits light inside a parallel sided glass sheet in air.

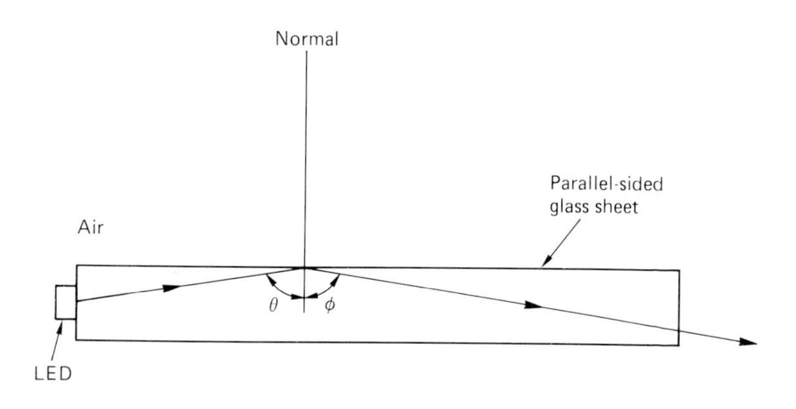

Figure A2.4 How light is transmitted along the inside of glass when the angle of incidence of the light is greater than the critical angle for the glass-air boundary

Light strikes the inside surface of the glass at an angle θ which is greater than the critical angle, so the light is internally reflected between the two glass-air boundaries along the glass sheet.

Optical fibres operate in this way, except that instead of glass-air boundaries, glass-glass boundaries are used, with each glass layer having a different refractive index. Figure A2.5 shows a cross-section of a circular optical fibre strand, with an internal cylindrical core of glass with a refractive index of, say, 1.5, surrounded by a glass cladding with an index of refraction of 1.4. The critical angle θ is calculated as:

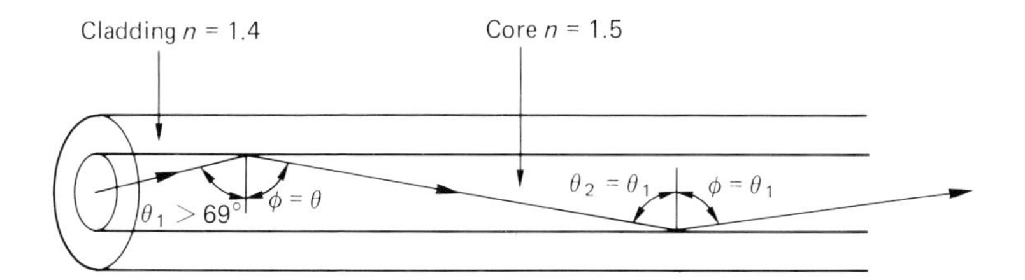

Figure A2.5 Typical construction of an optical fibre

$$\sin\theta = \frac{1.4}{1.5}$$

so θ is 69°. All light rays entering the fibre at angles over 69°, therefore, are internally reflected, remaining inside till they exit at the other end.

Optical fibre types

There are two main categories of optical fibre: **multimode** and **monomode**. In turn there are two main types of multimode fibre: **stepped-index** and **graded-index**. Multimode fibre is so called because there are many paths, called **modes**, which light may follow inside the fibre. Cross-sections of all three fibre types are shown in Figure A2.6.

A stepped-index multimode fibre is shown in Figure A2.6a alongside its **refractive index profile**, a simple graphical method of illustrating the refractive indices of the two glass layers of a fibre. The abrupt step between the refractive indices means that some light modes are longer than others, and so different, but simultaneously produced, light rays travelling along the fibre will arrive at the other end at different times. This **modal dispersion** means that the information the light carries is distorted – the amount of distortion increasing with fibre length. A limit of around 10 km is usual with stepped-index multimode fibre, due to modal disperson.

Graded-index multimode fibre (Figure A2.6b) reduces modal disperson with its varying core index of refraction. As the index varies so does the speed of light travelling through it – light in the centre of the core travels more slowly than light closer to the cladding. Overall effect is that light in all modes takes about the same time to be transmitted along the fibre. Modal dispersion is reduced and fibre length limit is increased to around 50 km.

Monomode fibre (A2.6c) has such a small diameter core that light may only travel along it in a single mode. Distortion due to different modal lengths is therefore eliminated and fibre lengths of hundreds of kilometres are possible.

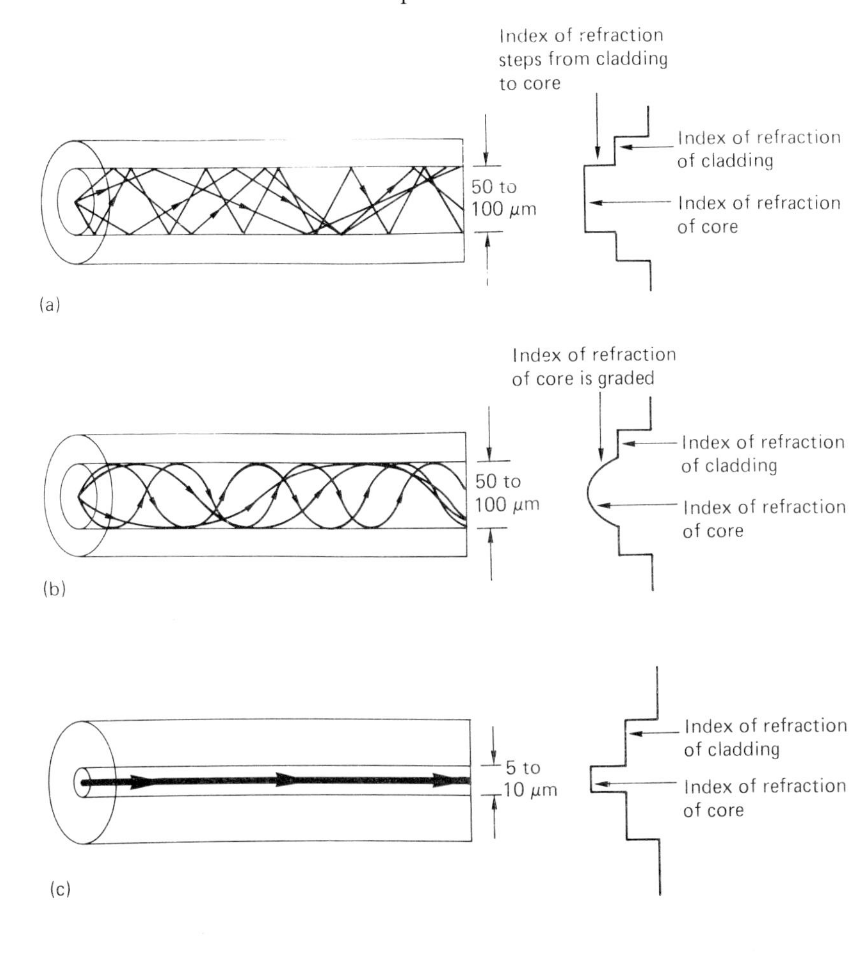

Figure A2.6 Cross-sections of the three main types of optical fibre, together with refractive index profiles (a) stepped-index multimode fibre (b) graded-index multimode fibre (c) monomode fibre

Attenuation

Loss occurs in an optical fibre mainly due to light being refracted or absorbed in the very glass itself by its intrinsic molecular nature. Generally, therefore, loss can only be minimised by reducing impurities within the glass.

Where loss is due to internal refraction, the process is known as **Rayleigh scattering** and is dependent, too, on light wavelength. In high-quality optical fibres, Rayleigh scattering causes losses of less than 1 dB km^{-1} at wavelengths around 0.9 μm and less than 0.5 dB km^{-1} at wavelengths around 1.3 μm. Although at first sight Rayleigh scattering is unwanted if we wish to have a fibre with low attenuation, it is useful in the principle of the optical time domain reflectometer, which measures the amount of Rayleigh scattered light returning to the equipment as backscatter. This is then translated by the optical time domain reflectometer into a display of optical fibre faults and distances.